全国高等中医药院校中药学类专业双语规划教材

Bilingual Planned Textbooks for Chinese Materia Medica Majors in TCM Colleges and Universities

无机化学实验

Inorganic Chemistry Experiment

（供中药学类、药学类及相关专业使用）
(For Chinese Materia Medica, Pharmacy and other related majors)

主　审　铁步荣

主　编　关　君　杨爱红

副主编　姚　军　马鸿雁　徐　旸　张晓青

编　者　（以姓氏笔画为序）

马鸿雁（成都中医药大学）	王　颖（北京中医药大学）
王　霞（河南中医药大学）	方德宇（辽宁中医药大学）
吕惠卿（浙江中医药大学）	齐学洁（天津中医药大学）
关　君（北京中医药大学）	杨　婕（江西中医药大学）
杨爱红（天津中医药大学）	李德慧（长春中医药大学）
吴巧凤（浙江中医药大学）	吴品昌（辽宁中医药大学）
张　璐（北京中医药大学）	张凤玲（浙江中医药大学）
张晓青（湖南中医药大学）	阿合买提江·吐尔逊（新疆医科大学）
林　舒（福建中医药大学）	罗　黎（山东中医药大学）
姚　军（新疆医科大学）	姚华刚（广东药科大学）
袁　洁（新疆医科大学厚博学院）	贾力维（黑龙江中医药大学）
徐　飞（南京中医药大学）	徐　旸（黑龙江中医药大学）
郭爱玲（山西中医药大学）	曹　莉（湖北中医药大学）
戴红霞（甘肃中医药大学）	

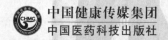

中国健康传媒集团

中国医药科技出版社

内 容 提 要

本教材是《全国高等中医药院校中药学类专业双语规划教材》之一，根据无机化学实验教学大纲的基本要求和课程特点编写而成。内容上涵盖了无机化学实验的基本理论知识、基本操作技能、化合物的提纯和制备、常数的测定以及一些综合性实验。本教材共选编了15个实验，每个实验均包括实验目的，实验原理，仪器、试剂及其他，实验内容，注意事项和思考题六部分。本教材为书网融合教材，即纸质教材有机融合电子教材、教学配套资源、数字化教学服务（在线教学、在线作业、在线考试），使教学资源更加多样化、立体化。

本教材可供全国普通高等医药院校中药学类、药学类及相关专业教学使用，也可作为从事相关工作的企业、研究院等科研工作者的参考用书。

图书在版编目（CIP）数据

无机化学实验：汉英对照/关君，杨爱红主编.—北京：中国医药科技出版社，2020.7

全国高等中医药院校中药学类专业双语规划教材

ISBN 978–7–5214–1882–8

Ⅰ.①无⋯　Ⅱ.①关⋯②杨⋯　Ⅲ.①无机化学–化学实验–双语教学–中医学院–教材–汉、英　Ⅳ.①O61-33

中国版本图书馆CIP数据核字（2019）第287417号

美术编辑	陈君杞
版式设计	辰轩文化
出版	中国健康传媒集团｜中国医药科技出版社
地址	北京市海淀区文慧园北路甲22号
邮编	100082
电话	发行：010-62227427　邮购：010-62236938
网址	www.cmstp.com
规格	889×1194 mm $\frac{1}{16}$
印张	12 $\frac{1}{2}$
字数	313千字
版次	2020年7月第1版
印次	2020年7月第1次印刷
印刷	三河市万龙印装有限公司
经销	全国各地新华书店
书号	ISBN 978-7-5214-1882-8
定价	44.00元

版权所有　盗版必究

举报电话：010-62228771

本社图书如存在印装质量问题请与本社联系调换

获取新书信息、投稿、为图书纠错，请扫码联系我们。

出版说明

近些年随着世界范围的中医药热潮的涌动，来中国学习中医药学的留学生逐年增多，走出国门的中医药学人才也在增加。为了适应中医药国际交流与合作的需要，加快中医药国际化进程，提高来中国留学生和国际班学生的教学质量，满足双语教学的需要和中医药对外交流需求，培养优秀的国际化中医药人才，进一步推动中医药国际化进程，根据教育部、国家中医药管理局、国家药品监督管理局等部门的有关精神，在本套教材建设指导委员会主任委员成都中医药大学彭成教授等专家的指导和顶层设计下，中国医药科技出版社组织全国50余所高等中医药院校及附属医疗机构约420名专家、教师精心编撰了全国高等中医药院校中药学类专业双语规划教材，该套教材即将付梓出版。

本套教材共计23门，主要供全国高等中医药院校中药学类专业教学使用。本套教材定位清晰、特色鲜明，主要体现在以下方面。

一、立足双语教学实际，培养复合应用型人才

本套教材以高校双语教学课程建设要求为依据，以满足国内医药院校开展留学生教学和双语教学的需求为目标，突出中医药文化特色鲜明、中医药专业术语规范的特点，注重培养中医药技能、反映中医药传承和现代研究成果，旨在优化教育质量，培养优秀的国际化中医药人才，推进中医药对外交流。

本套教材建设围绕目前中医药院校本科教育教学改革方向对教材体系进行科学规划、合理设计，坚持以培养创新型和复合型人才为宗旨，以社会需求为导向，以培养适应中药开发、利用、管理、服务等各个领域需求的高素质应用型人才为目标的教材建设思路与原则。

二、遵循教材编写规律，整体优化，紧跟学科发展步伐

本套教材的编写遵循"三基、五性、三特定"的教材编写规律；以"必需、够用"为度；坚持与时俱进，注意吸收新技术和新方法，适当拓展知识面，为学生后续发展奠定必要的基础。实验教材密切结合主干教材内容，体现理实一体，注重培养学生实践技能训练的同时，按照教育部相关精神，增加设计性实验部分，以现实问题作为驱动力来培养学生自主获取和应用新知识的能力，从而培养学生独立思考能力、实验设计能力、实践操作能力和可持续发展能力，满足培养应用型和复合型人才的要求。强调全套教材内容的整体优化，并注重不同教材内容的联系与衔接，避免遗漏和不必要的交叉重复。

三、对接职业资格考试,"教考""理实"密切融合

本套教材的内容和结构设计紧密对接国家执业中药师职业资格考试大纲要求,实现教学与考试、理论与实践的密切融合,并且在教材编写过程中,吸收具有丰富实践经验的企业人员参与教材的编写,确保教材的内容密切结合应用,更加体现高等教育的实践性和开放性,为学生参加考试和实践工作打下坚实基础。

四、创新教材呈现形式,书网融合,使教与学更便捷更轻松

全套教材为书网融合教材,即纸质教材与数字教材、配套教学资源、题库系统、数字化教学服务有机融合。通过"一书一码"的强关联,为读者提供全免费增值服务。按教材封底的提示激活教材后,读者可通过PC、手机阅读电子教材和配套课程资源(PPT、微课、视频等),并可在线进行同步练习,实时收到答案反馈和解析。同时,读者也可以直接扫描书中二维码,阅读与教材内容关联的课程资源,从而丰富学习体验,使学习更便捷。教师可通过PC在线创建课程,与学生互动,开展在线课程内容定制、布置和批改作业、在线组织考试、讨论与答疑等教学活动,学生通过PC、手机均可实现在线作业、在线考试,提升学习效率,使教与学更轻松。此外,平台尚有数据分析、教学诊断等功能,可为教学研究与管理提供技术和数据支撑。需要特殊说明的是,有些专业基础课程,例如《药理学》等9种教材,起源于西方医学,因篇幅所限,在本次双语教材建设中纸质教材以英语为主,仅将专业词汇对照了中文翻译,同时在中国医药科技出版社数字平台"医药大学堂"上配套了中文电子教材供学生学习参考。

编写出版本套高质量教材,得到了全国知名专家的精心指导和各有关院校领导与编者的大力支持,在此一并表示衷心感谢。希望广大师生在教学中积极使用本套教材和提出宝贵意见,以便修订完善,共同打造精品教材,为促进我国高等中医药院校中药学类专业教育教学改革和人才培养做出积极贡献。

全国高等中医药院校中药学类专业双语规划教材建设指导委员会

主 任 委 员 彭 成（成都中医药大学）

副主任委员 （以姓氏笔画为序）

　　　　　　　朱卫丰（江西中医药大学）　　闫永红（北京中医药大学）
　　　　　　　邱　峰（天津中医药大学）　　邱智东（长春中医药大学）
　　　　　　　胡立宏（南京中医药大学）　　容　蓉（山东中医药大学）
　　　　　　　彭代银（安徽中医药大学）

委　　　员 （以姓氏笔画为序）

　　　　　　　王小平（陕西中医药大学）　　王光志（成都中医药大学）
　　　　　　　韦国兵（江西中医药大学）　　邓海山（南京中医药大学）
　　　　　　　叶耀辉（江西中医药大学）　　刚　晶（辽宁中医药大学）
　　　　　　　刘中秋（广州中医药大学）　　关　君（北京中医药大学）
　　　　　　　杨光明（南京中医药大学）　　杨爱红（天津中医药大学）
　　　　　　　李　楠（成都中医药大学）　　李小芳（成都中医药大学）
　　　　　　　吴锦忠（福建中医药大学）　　张　梅（成都中医药大学）
　　　　　　　张一昕（河北中医学院）　　　陆兔林（南京中医药大学）
　　　　　　　陈胡兰（成都中医药大学）　　邵江娟（南京中医药大学）
　　　　　　　周玖瑶（广州中医药大学）　　赵　骏（天津中医药大学）
　　　　　　　胡冬华（长春中医药大学）　　钟凌云（江西中医药大学）
　　　　　　　侯俊玲（北京中医药大学）　　都晓伟（黑龙江中医药大学）
　　　　　　　徐海波（成都中医药大学）　　高增平（北京中医药大学）
　　　　　　　高德民（山东中医药大学）　　唐民科（北京中医药大学）
　　　　　　　寇晓娣（天津中医药大学）　　蒋桂华（成都中医药大学）
　　　　　　　韩　丽（成都中医药大学）　　傅超美（成都中医药大学）

数字化教材编委会

主　审　铁步荣

主　编　关　君　杨爱红

副主编　姚　军　马鸿雁　徐　旸　张晓青

编　者（以姓氏笔画为序）

马鸿雁（成都中医药大学）　　　　　王　颖（北京中医药大学）
王　霞（河南中医药大学）　　　　　方德宇（辽宁中医药大学）
吕惠卿（浙江中医药大学）　　　　　齐学洁（天津中医药大学）
关　君（北京中医药大学）　　　　　杨　婕（江西中医药大学）
杨爱红（天津中医药大学）　　　　　李德慧（长春中医药大学）
吴巧凤（浙江中医药大学）　　　　　吴品昌（辽宁中医药大学）
张　璐（北京中医药大学）　　　　　张凤玲（浙江中医药大学）
张晓青（湖南中医药大学）　　　　　阿合买提江·吐尔逊（新疆医科大学）
林　舒（福建中医药大学）　　　　　罗　黎（山东中医药大学）
姚　军（新疆医科大学）　　　　　　姚华刚（广东药科大学）
袁　洁（新疆医科大学厚博学院）　　贾力维（黑龙江中医药大学）
徐　飞（南京中医药大学）　　　　　徐　旸（黑龙江中医药大学）
郭爱玲（山西中医药大学）　　　　　曹　莉（湖北中医药大学）
戴红霞（甘肃中医药大学）

前 言

为了满足21世纪对国际化高素质人才的迫切需求、推进高等教育的内涵式发展和高等院校的"双一流"建设，我们组织编写了这本供中药学类本科各专业使用的《无机化学实验》双语教材。

本教材由国内18所医药院校的28名一线教师通力合作，经过多次集体研究讨论，分工编写，精心修改后由主编统稿完成。在编写过程中，本教材编委会充分调研、总结了近十年来各个高校在无机化学双语教学课程体系、教学内容和实验教学方面所取得的成果，吸取了国内外化学和药学相关教材编写的经验。

本教材全书分为三章，其中第一、二章没有具体实验，但它们是无机化学实验课的基础，学生在进入实验室前应该先学习这两章的内容。第三章包括15个实验，每个实验均包括实验目的，实验原理，仪器、试剂及其他，实验内容，注意事项和思考题等部分。除验证性实验之外，本教材还特别编写了综合实验和设计实验，目的在于培养学生独立进行实验的能力，让学生可以选择适当的题目，自行设计实验方案和步骤，在课堂或在开放实验室中独立完成实验，并写出实验报告。本教材所有实验都是来自各参编学校目前实验教学的内容。同时也囊括了其他院校实验课的一些内容。由于各校所开实验不完全相同，因此所编写的实验比实际课时所允许的要多一些，可根据实际情况选用。

本教材编写分工如下：第一章绪论由杨爱红、齐学洁、王霞、阿合买提江·吐尔逊、罗黎负责；第二章常用仪器与基本操作由关君、贾力维、王颖、张璐、吴品昌负责；第三章实验项目分工是：杨婕（实验一）；吕惠卿（实验二）；马鸿雁（实验三）；曹莉（实验四）；吴巧凤、张凤玲（实验五）；姚军（实验六）；袁洁（实验七）；林舒（实验八）；徐飞（实验九）；戴红霞（实验十）；徐旸（实验十一）；方德宇（实验十二）；张晓青（实验十三）；姚华刚（实验十四）；李德慧（实验十五）；郭爱玲（附录）。全书由关君、杨爱红负责统稿定稿，铁步荣负责主审。

本教材可作为普通高等医药院校中药学类、药学类及相关专业无机化学实验课程的教科书，也可作为化学、医学、环境等相关专业无机化学实验课程的教学参考书，还可作为科研院所、医药企业、药品管理机构等的参考书。

本教材的编写工作获得了全体编者所在院校的大力支持与帮助，铁步荣教授对本教材的审定工作给予了极大的帮助，在此一并致谢。

由于编者学识水平所限，本教材可能会存在不妥与疏漏之处，敬请广大师生和读者提出宝贵意见，以便再版时加以修改和完善。

编 者
2020年6月

Preface

In order to meet the urgent needs of international high-quality talents in the 21st century, and to promote the connotative development of higher education and "double first class" construction of colleges and universities, according to the spirits of several meetings of the editorial board of bilingual planning teaching materials of traditional Chinese medicine major in national colleges and universities of traditional Chinese Medicine, we organized and compiled this bilingual textbook of inorganic chemistry experiment for Chinese medicine undergraduates.

This textbook is finished by 28 front-line teachers from 18 medical universities in China. After many collective research and discussion, division of labour, compiling, and careful revision, the manuscript was unified by the chief editor. In the process of compiling, the editorial board of this textbook fully investigated and summarized the achievements of each university in the course system, teaching content and experimental teaching of inorganic chemistry bilingual teaching in the past ten years, and drew on the experience of compiling chemistry and pharmacy related textbooks at home and abroad.

This textbook is divided into three chapters. There are no specific experiments in Chapters 1 and 2, but they are the basis of inorganic chemistry experiment. Students should learn the contents of these two chapters before entering the laboratory. There are 15 experiments in Chapter 3, each experiment consists of objectives; principles; instrument, reagent and others; experiment content; notes; and questions. In addition to the confirmatory experiments, the textbook also specially compiles the comprehensive experiments and designing experiments, aiming to cultivate the ability of students to carry out the experiment independently. So that students can independently choose appropriate topic, design the experiment scheme and steps, complete the experiment in the classroom or in the open laboratory, and write the experiment report all by themselves. All experiments in this textbook are from the current experimental teaching content of participating universities. At the same time, some contents of experimental courses in other universities are also absorbed. Because the experiments in different universities are not entirely the same, so the experiments written are more than those allowed in the actual class hours, which can be selected according to the actual situation.

The division of labor in the compilation of this textbook is as follows: the first chapter is in the charge of Aihong Yang, Xuejie Qi, Xia Wang, Ahmatjian Tursun and Li Luo; the second chapter is in the charge of Jun Guan, Liwei Jia, Ying Wang, Lu Zhang and Pinchang Wu; the third chapter is in the charge of Jie Yang (Experiment 1); Huiqing Lv (Experiment 2); Hongyan Ma (Experiment 3); Li Cao (Experiment 4); Qiaofeng Wu and Fengling Zhang (Experiment 5); Jun Yao (Experiment 6); Jie Yuan (Experiment 7); Shu Lin (Experiment 8); Fei Xu (Experiment 9); Hongxia Dai (Experiment 10); Yang Xu (Experiment 11); Deyu Fang (Experiment 12); Xiaoqing Zhang (Experiment 13); Huagang Yao (Experiment 14); Dehui Li (Experiment 15); Ailing Guo (Appendix). Jun Guan and Airong Yang are in charge of the finalization of the whole book, while Burong Tie is in charge of the review.

This textbook can be used as inorganic chemistry experiment textbook for Chinese Materia Medica, Pharmacy and other relative majors in TCM colleges and universities, and as a teaching reference book for inorganic chemistry experiment for chemistry, medicine, environment and other majors, and also as a professional reference book for scientific research institutes, pharmaceutical enterprises and pharmaceutical management organizations, etc.

When compiling this textbook, we were greatly supported and helped by the universities where the editorial committee come from. In particular, We recieved great help from Proffessor Burong Tie in examining and approving the textbook. We therefore express our sincere thanks to them.

There might still exist something improper and inadvertent in this textbook due to our academic limitations. We would be most appreciative if teachers, students and readers could give us valuable suggestions on improving this textbook, so that we will revise and improve it in the next edition.

Authors
June 2020

目录 | Contents

第一章　绪论 ·· 1
Chapter 1　**Introduction** ·· 10

第二章　常用仪器与基本操作 ·· 22
Chapter 2　**Common Instruments and Basic Operation** ·· 49

第三章　实验项目 ·· 86
　实验一　仪器的认领和基本操作训练 ·· 86
Chapter 3　**Experiment Items** ·· 88
　Experiment 1　Cognition and Basic Operation Training of Instruments ······················· 88

　实验二　电解质溶液 ··· 91
　Experiment 2　Electrolyte Solution ··· 94

　实验三　缓冲溶液的配制与性质 ·· 98
　Experiment 3　Preparation and Properties of Buffer Solution ···································· 101

　实验四　醋酸电离度和电离平衡常数的测定 ·· 105
　Experiment 4　Determination of Ionization Degree and Equilibrium Constant of
　　　　　　　　　Acetic Acid ··· 107

　实验五　氯化铅溶度积常数的测定 ·· 110
　Experiment 5　Determination of Solubility Product Constant of $PbCl_2$ ···················· 112

　实验六　氧化还原反应与电极电势 ·· 115
　Experiment 6　Redox Reaction and Electrode Potential ·· 118

　实验七　配合物的生成、性质与应用 ·· 122
　Experiment 7　Formation, Properties and Application of Complexes ························· 125

　实验八　银氨配离子配位数的测定 ·· 129
　Experiment 8　Determining Coordination Number of $[Ag(NH_3)_n]^+$ ······················ 132

　实验九　铬、锰、铁 ··· 135
　Experiment 9　Chromium, Manganese and Iron ··· 138

　实验十　药用氯化钠的制备 ··· 142
　Experiment 10　Preparation of Medicinal Sodium Chloride ······································ 144

　实验十一　药用氯化钠的性质及杂质限量的检查 ··· 146
　Experiment 11　Inspection of Properties and Impurity Limits of Medicinal
　　　　　　　　　　Sodium Chloride ·· 149

1

实验十二　硫酸亚铁铵的制备 ··· 153
Experiment 12　Preparation of Ammonium Ferrous Sulfate ·· 155

实验十三　葡萄糖酸锌的制备 ·· 158
Experiment 13　Preparation of Zinc Gluconate ·· 160

实验十四　三草酸合铁（Ⅲ）酸钾的制备和性质 ·· 162
Experiment 14　Preparation and Properties of Potassium Trioxalatoferrate（Ⅲ）Trihydrate ······ 167

实验十五　无机阴、阳离子的鉴定和矿物药的鉴别 ··· 173
Experiment 15　Identification of Inorganic Anions, Cations, and Mineral Drugs ··················· 176

附录 ·· 180
附录1　实验室常用试剂的配制 ·· 180
附录2　常用的酸碱指示剂 ·· 183
附录3　常见离子和化合物的颜色 ··· 184
附录4　常见阴、阳离子鉴定一览表 ·· 186

参考文献 ·· 188

第一章 绪 论

一、无机化学实验教学的目的与要求

化学是一门以实验为基础的自然科学，尽管现代科学技术飞速发展，现代化学已经进入了理论与实践并重的阶段，但实验有其不可替代的作用，它始终是检验理论是否正确的唯一标准。化学实验课是实施全面化学教育的一种最有效的形式，在实验课中学生是学习的主体，在教师的指导下，学生自己动手进行实验操作，观察记录实验现象，处理实验数据，撰写实验报告，自己动脑筋解决各种各样的问题，各项智力因素都能得到发展。

无机化学是药学类学生所学的第一门化学基础课，要掌握无机化学的基本知识和基本理论，无机化学实验教学是必不可少的重要环节。由于无机化学实验是在一年级开设，一年级是学生从高中向大学过渡的重要阶段，因此教师要充分发挥指导作用，调动学生的积极能动性。教与学双方均应积极努力，要对实验课给予足够重视。

1. 无机化学实验教学的目的

（1）学生在无机化学实验中可获得大量的感性知识，通过归纳、总结由感性认识上升到理性认识，对课堂讲授的基本理论和基础知识的理解和掌握会更加深刻。通过实验，使理论与实践相结合，培养学生理论联系实际的能力，以及分析问题解决问题的能力。

（2）通过无机化学实验基本操作技能的训练，使学生掌握无机化学实验基本操作方法和技能技巧，掌握常见元素及其化合物的重要性质和反应规律，了解无机化合物的一般提纯和制备方法，为后续各门课程的实验打下坚实的基础。

（3）培养学生独立进行实验的能力、观察和记录实验现象的能力、正确处理实验数据的能力和撰写实验报告的能力。

（4）培养学生严格的"数"与"量"的概念，使学生树立严谨的科学态度。

（5）在实验中逐步培养准确细致、认真整洁地进行科学实验的良好习惯。

（6）通过实验课程的学习，使学生受到规范、系统的无机化学实验训练。逐步培养学生科学的逻辑思维方法；严谨认真、实事求是、一丝不苟的科学态度和求真、探索、协作的精神；以及创新思维和创新实践的能力。

实验是全面提高学生综合能力的重要环节，它既有对知识的传授，又有操作技能技巧的训练；既有逻辑思维方法的训练，又有良好的工作习惯、作风和科学工作方法的培养，因此无机化学实验课程在后续专业实验课程的学习中具有重要的意义。

2. 无机化学实验教学的要求

（1）教师要认真、负责、严格要求学生，特别要重视对学生基本操作技能的训练和实验工作能力及良好的工作习惯的培养，并贯穿在各个具体的实验环节中。

（2）教师要坚持在每次实验前对预习情况进行提问，讲解重点内容；实验中认真巡视指导，解惑答疑，及早发现问题、解决问题；实验后进行总结、讨论，优化教学方法；课下认真批改实验报告。

（3）学生要明确低年级基本操作技能的训练和实验能力的培养，是高年级各门实验课程甚至是以后从事科学研究和实际工作必备的基础。

（4）学生从实验的各个环节都要对自己严格要求，对每一个实验，不仅要理解掌握实验的原理、实验的方法，而且要对基本操作进行认真的训练，对每一个操作都要进行一丝不苟的练习。

（5）学生在实验过程中应该养成良好的工作习惯和工作作风，比如实验台面的整洁、仪器存放的有序、污物不乱丢弃、实验试剂的定点回收等。

（6）要重视实验中的每个环节，尤其是实验室的安全操作规范一定要熟知。

二、无机化学实验的学习方法

要达到以上的学习目的，不仅要有正确的学习态度，还需要有正确的学习方法。做好无机化学实验必须掌握以下三个重要步骤。

第一步：预习。实验前的预习是保证做好实验的一个非常重要的环节，通过预习应达到以下要求。

（1）认真阅读实验教材、观看相关实验操作的教学视频资料，注重实验教材和理论教材的相互联系，运用理论指导实验。

（2）明确实验的目的。

（3）熟悉实验内容、基本原理、实验步骤、基本操作和注意事项。

（4）认真思考实验教材中的思考题，对可能出现的实验现象作出自己的推断和解释，对于自己不能解决的问题也可通过小组讨论的方式得出答案。

（5）写好预习报告。

第二步：实验。实验过程中学生应遵守实验室规则，接受教师指导，根据实验教材和标准化操作的教学视频资料上所规定的方法、步骤和试剂用量严格进行操作，并应做到以下几点。

（1）认真操作，细心观察，严格按照各项实验的基本操作规程进行，如实记录实验中观察到的现象和实验数据。

（2）如果得到的实验现象或结果和预期不符合，应认真分析其原因，及时纠正实验错误，并考虑是否需要重做实验。

（3）实验中遇到疑难问题或突发事件时，要及时向教师汇报。

（4）在实验过程中应该保持肃静，严格遵守实验室工作规则。

第三步：总结。实验结束后，应根据实验现象给出实验结论，或者根据实验数据进行处理和计算，独立完成实验报告，并及时总结自己在实验中的收获和不足，然后交给指导教师评阅。

（1）若实验失败，要及时总结实验失败的原因，找出实验过程中的操作错误和缺点，以免类似问题的再次出现。

（2）若有实验现象、解释、结论、数据等不符合要求，应考虑重做实验；若有抄袭实验报告、篡改实验数据、字迹潦草者应重写报告。

（3）实验报告书写时应字迹端正，简明扼要，整齐清洁。

三、实验室规则

（1）实验室是教学和科研的场所，一般不作他用。

（2）进入实验室的一切人员必须遵守纪律，保持肃静，集中思想，认真操作，仔细观察，积极思考，真实记录。

（3）实验前必须预习实验内容，明确目的要求，熟悉方法步骤，掌握基本原理。

（4）正确使用实验仪器、设备。使用精密仪器时，必须严格按照操作规程进行操作，细心谨慎，避免粗枝大叶而损坏仪器。如发现仪器有故障，应立即停止使用并报告指导教师，及时排除故障。

（5）药品应按规定的量取用，已取出的试剂不能再放回原试剂瓶中以免污染试剂。取用药品的用具应保持清洁、干燥，以保证试剂不被污染及浓度一定。取用药品后立即盖上瓶盖，以免放错瓶塞，污染药品。

（6）实验前要检查所需仪器是否齐全，是否破损，以便及时补齐、更换。实验过程中应保持器皿清洁，保持实验台面清洁整齐，实验结束后仪器、药品放回原处。

（7）废的固体、纸、玻璃碴、火柴梗等应倒入废品篮内；废液应倒入指定的废液回收桶，不得倒入水槽流入下水道，剧毒废液由实验室统一处理。

（8）实验完成后应保持实验室清洁，检查水、电、气安全，关好门窗。

（9）实验室内一切物品（仪器、药品和产物等）不得带离实验室。离开实验室前，需经指导教师签字。

（10）如果发生意外事故，应保持镇静，不要惊慌失措；偶有烧伤、烫伤、割伤应及时报告教师，进行急救和治疗。

四、实验室安全守则

无机化学实验用到的药品中有的是易燃、易爆品，有的具有腐蚀性和毒性。因此，实验中要特别注意安全，必须将"安全"放在首位。

（1）进入实验室前，应首先查看灭火设施的存放位置及熟悉应急通道，学会消防器材的使用方法。

（2）金属钾、钠和白磷等暴露在空气中易燃烧，所以金属钾、钠应保存在煤油中，白磷则可保存在水中。取用它们时要用镊子。有一些有机溶剂（如乙醚、乙醇、丙酮、苯等）极易燃，使用时必须远离明火、热源，用毕立刻盖紧瓶塞，存放在安全的地方。

（3）含氧的气体遇火易爆炸，操作时必须严禁接近明火。点燃氢气前，必须先检查并确保纯度。银氨溶液不能留存，因久置后会变成易爆炸的物质。某些强氧化剂（如氯酸钾、硝酸钾、高锰酸钾等）或其混合物不能研磨，否则将引起爆炸。

（4）配备必要的护目镜。倾注药剂或加热液体时，容易溅出，不要俯视容器。尤其是浓酸、浓碱具有强腐蚀性，切勿使其溅在皮肤或衣服上，眼睛应注意保护。稀释它们时（特别是浓硫酸），应将它们慢慢倒入水中，而不能相反进行，以避免溅出。加热试管时，切记不要使试管口向着自己或别人。

（5）产生有刺激性气味、有毒气体的实验要在通风橱中进行，嗅气体的气味时，只能用手轻轻地扇动空气，使少量气体进入鼻孔。

（6）使用有毒试剂如砷化物、铬盐、氰化物、汞及其化合物等，要严格防止进入口内或伤口内，废液严禁排入下水道。

（7）金属汞易挥发，并通过呼吸道进入人体内，逐渐积累会引起慢性中毒，所以金属汞的实验应特别小心，不得把金属汞洒落在桌上或地上。一旦洒落，必须尽可能收集起来，并用硫黄粉盖在洒落的地方，使金属汞转变成不挥发难溶的硫化汞。

（8）用指定的药匙或容器取用药品，决不允许品尝药品。不允许随便将各种药品混合，以免发生意外事故。自行设计的实验须征得教师同意后方可进行。

（9）湿手不要接触电器插头，人体不能与导电物体接触。实验结束后应切断电源。

（10）严禁在实验室内饮食、吸烟、不得将食物或餐饮具带入实验室，实验后要清洗双手。

五、实验室事故的处理

化学实验室存有大量易燃易爆和有毒有害化学药品,因此进行化学实验时要做好防火、防爆和防触电工作。

1. 防火

(1)操作时注意(远离火源)切忌将易燃溶剂放在广口容器直火加热;加热必须在水浴中进行,切忌当附近有暴露的易燃溶剂时点火。

(2)进行易燃物质实验时,应当养成先将乙醇等易燃物质移开的习惯。

(3)蒸馏易燃的有机物时,装置不能漏气,如发生漏气时,应立即停止加热,检查原因。

(4)使用大量易燃液体时,应在通风橱内或指定的地方进行,室内应无火源。

(5)不得把燃烧或带有火星的火柴杆或纸条等乱抛乱扔。

(6)直火加热时,实验者不得擅自离开实验室。

2. 防爆

(1)易燃易爆溶剂(如乙醚、汽油等)切勿接近火源。

(2)不能重压或撞击易爆炸固体(如重金属、乙炔化物、苦味酸金属盐、三硝基甲苯等)。

3. 防触电

(1)使用电器前,先了解电器对电源的要求及匹配,选择好相应的插座或导线。

(2)使用时必须检查好线路再插上电源,实验结束时必须先切断电源再拆线路。

4. 处理方法   如果在实验过程中不慎发生意外事故,请不要慌张,应沉着、冷静,迅速做出处理。

(1)烫伤   轻微烫伤可先用清水冲洗,再搽上烫伤油膏。如果烫伤较重,应立即到医务室医治。

(2)割伤   伤处不能用手抚摸,也不能用水洗涤。若是玻璃创伤,应先把碎玻璃从伤处挑出,再在伤口处涂抹紫药水或红药水,必要时撒些消炎粉或敷些消炎膏,再用纱布包扎。

(3)眼睛受伤   防止眼睛受刺激性气体熏染,防止任何化学试剂特别是强酸、强碱、玻璃屑等异物进入眼内,特殊实验要佩戴护目镜。一旦眼内溅入任何化学试剂,立即用大量水缓缓彻底冲洗15分钟,将伤者送医院治疗。玻璃屑进入眼内时,要尽量保持平静,绝不可用手揉搓,也不要试图让别人取出碎屑,任其流泪,用纱布包住伤者眼睛后,急送医院处理。

(4)酸腐蚀   发生在皮肤上,用大量水冲洗,用5%碳酸氢钠溶液洗涤,再涂油膏;在眼睛上,抹去溅在眼睛外面的酸,立即用水冲洗,用洗眼器对准眼睛冲洗,再用稀碳酸氢钠洗涤,最后滴少量麻油;发生在衣服上,先用水冲洗,再用稀碱水洗,最后用水重新清洗。

(5)碱腐蚀   发生在皮肤上,用饱和硼酸溶液或1%醋酸溶液洗涤,再涂上油膏;在眼睛上,擦去眼睛外的碱,用水洗,再用稀酸中和多余的碱,再用水冲洗。

(6)火灾处理   首先切断电源,关闭煤气,搬开易燃物品;电话报警;对可溶于水的液体着火时,可用湿布或水灭火;对密度小于水的非水溶性的有机试剂着火时,用砂土灭火(不可用水);如火势较大,可使用CCl_4灭火器或CO_2泡沫灭火器,但不可用水扑救,因水能和某些化学药品(如金属钠)发生剧烈的反应而引起更大的火灾。如遇电气设备着火,必须使用CCl_4灭火器,绝对不能用水或CO_2泡沫灭火器。

(7)触电   遇有触电事故,首先应切断电源,然后在必要时,进行人工呼吸。

(8)伤势较重者,立即送医。

六、实验室"三废"的处理

化学实验中经常会产生有毒的气体、液体和固体，都需要及时处理，特别是某些剧毒物质，如果直接排出就可能污染周围空气和水源，损害人体健康。因此，对废液和废气、废渣要经过一定的科学处理后，才能排弃。

1. 废气的处理 产生有毒气体的实验应在通风橱内进行。少量有毒气体可以通过排风设备排出室外，被空气稀释。毒气量大时，实验必须备有吸收或处理装置，经处理后再排出。如二氧化氮、二氧化硫、氯气、硫化氢、氟化氢等可用导管通入碱液中，使其大部分吸收后排出，一氧化碳可点燃转成二氧化碳。

2. 废液的处理

（1）化学实验中大量的废液通常是废酸液。废酸液可先用耐酸塑料网纱或玻璃纤维过滤，滤液加碱中和，调节 pH 至 6~8 后就可排出。少量滤渣可埋于地下。

（2）铬酸洗液的废液可以用高锰酸钾氧化法使其再生，重复使用。氧化方法：先在 110~130℃ 下将其不断搅拌、加热、浓缩，除去水分后，冷却至室温，缓缓加入高锰酸钾粉末。每 1000ml 加入 10g 左右，边加边搅拌直至溶液呈深褐色或微紫色，不要过量。然后直接加热至有三氧化硫出现，停止加热。稍冷，通过玻璃砂芯漏斗过滤，除去沉淀；冷却后析出红色三氧化铬沉淀，再加适量硫酸使其溶解即可使用。少量的铬酸洗液废液可加入废碱液或石灰使其生成氢氧化铬（Ⅲ）沉淀，将此废渣埋于地下。

（3）氰化物是剧毒物质，含氰废液必须认真处理。含氰化物的废液用氢氧化钠溶液调至 pH 10 以上，再加入 3% 的高锰酸钾使 CN^- 氧化分解。CN^- 含量高的废液用碱性氯化法处理，即在 pH 10 以上加入次氯酸钠使 CN^- 氧化分解。

（4）含汞盐的废液先调至 pH 8~10，加入过量硫化钠，使其生成硫化汞沉淀，再加入共沉淀剂硫酸亚铁，生成的硫化铁将水中的悬浮物硫化汞微粒吸附而共沉淀，排出清液，残渣用焙烧法回收汞或再制成汞盐。清液汞含量降到 $0.02mg \cdot L^{-1}$ 以下可排放。少量残渣可埋于地下，大量残渣可用焙烧法回收汞；但要注意一定要在通风橱内进行。

（5）含砷废液加入氧化钙，调节 pH 为 8，生成砷酸钙和亚砷酸钙沉淀。或调节 pH 10 以上，加入硫化钠与砷反应，生成难熔、低毒的硫化物沉淀。

（6）含铅、镉废液，用消石灰将 pH 调至 8~10，使 Pb^{2+}、Cd^{2+} 生成 $Pb(OH)_2$ 和 $Cd(OH)_2$ 沉淀，加入硫酸亚铁作为共沉淀剂。

（7）含重金属离子的废液，最有效和最经济的处理方法是加碱或加硫化钠把重金属离子变成难溶性的氢氧化物或硫化物沉积下来，然后过滤分离，少量残渣可埋于地下。

（8）低浓度含酚废液加次氯酸钠或漂白粉使酚氧化为二氧化碳和水。高浓度含酚废水用乙酸丁酯萃取，重蒸馏回收酚。

（9）有机溶剂 可燃性有机废液可于燃烧炉中通氧气完全燃烧。一些废有机溶剂可以通过回收进行处理。

废乙醚溶液置于分液漏斗中，用水洗一次，中和，用 0.5% 高锰酸钾洗至紫色不褪，再用水洗，用 0.5%~1% 硫酸亚铁铵溶液洗涤，除去过氧化物，再用水洗，用氯化钙干燥、过滤、分馏，收集 33.5~34.5℃ 馏分。

乙酸乙酯废液：先用水洗几次，再用硫代硫酸钠稀溶液洗几次，使之褪色，再用水洗几次，蒸馏，用无水碳酸钾脱水，放置几天，过滤后蒸馏，收集 76~77℃ 馏分。

三氯甲烷、乙醇、四氯化碳等废溶液都可以通过水洗废液再用试剂处理，最后通过蒸馏收集沸点左右馏分，得到可再用的溶剂。

3. 废料的处理 实验中出现的固体废弃物不能随便乱放，以免发生事故。如能放出有毒气体或能自燃的危险废料不能丢进废品箱内和排进废水管道中。不溶于水的废弃化学药品禁止丢进废水管道中，必须将其在适当的地方烧掉或用化学方法处理成无害物。碎玻璃和其他有棱角的锐利废料，不能丢进废纸篓内，要收集于特殊废品箱内处理。

七、实验数据处理

1. 关于实验记录

（1）实验记录本要逐页编号，不允许撕掉，原始记录一定要严格，养成保留原始记录的习惯。

（2）实验记录要规范，如有记录数据错误，不能完全涂掉或掩盖，应该能看到数据改动的痕迹，以便分析错误的原因。

（3）实验原始记录要求当场完整如实记录在记录本的左侧，处理数据结果写在右侧，切忌记录在纸上然后再誊抄到记录本上。

2. 实验数据统计与整理的一般方法

（1）数据的表格化

① 根据实验原理确定应记录的项目：所测的物理量。

② 注意数据的有效数字及单位和必要的注解。

③ 设计好的表格要便于数据的查找、比较，便于数据的计算和进一步处理，便于反映数据间的关系。

（2）数据的图像化　图像化是用直线图或曲线图对化学实验结果进行处理的一种简明化形式，适用于一个量的变化引起另一个量的变化的情况。图像化的最大特点是简洁直观。

3. 实验数据筛选与处理策略 对实验数据筛选的一般方法和思路为"五看"：一看数据是否符合测量仪器的精确度，如用托盘天平测得的质量精确度为0.1g，若精确度值超过了这个范围，说明所得数据无效；二看数据是否在误差允许范围内，若所得的数据明显超出误差允许范围，要舍去；三看反应是否完全，是否是不足量反应物作用下所得的数据，只有完全反应时所得的数据才能应用；四看所得数据的测量环境是否一致，特别是气体体积，只有在温度、压强一致的情况下才能进行比较、运算；五看数据测量过程是否规范、合理，错误和违反测量规则的数据要舍去。

4. 关于有效数字 在化学实验中，经常要根据实验测得的数据进行化学计算，为了取得准确的结果，不仅要准确进行测量，而且还要正确记录与计算。正确记录是指正确记录数字的位数，因为数据的位数不仅表示数字的大小，也反映测量的准确程度。

确定有效数字位数时应遵循以下几条原则：

（1）在记录测量数据时，只允许在测得值的末位保留一位估读数字。

（2）变换单位时，有效数字的位数必须保持不变。例如：0.0087g应写成8.7mg，23.6L应写成2.36×10^4ml。

（3）对于很大或很小的数字，可以用指数形式表示。如0.00025g可记录为2.5×10^{-4}g。又如2500g，若为3位有效数字，可记录为2.50×10^3g。

（4）对于pH及pK_a等对数值，其有效数字仅取决于小数部分数字的位数，而其整数部分的数值只代表原数值的幂次。例如：pH=10.14，对应的$[H^+]=7.2\times10^{-11}$mol·L^{-1}，有效数字是2位，而不是4位。

在数据处理过程中应注意以下方面：

（1）若测定结果是由几个测定值相加或相减所得，保留有效数字的位数取决于小数点后位数最少的一个；若测定结果是由几个测量值相除所得，则保留有效数字的位数取决于有效数字位数最少的一个。

（2）将多余的数字舍去，所采用的规则一般是"四舍六入五成双"。即当尾数≤4时舍去，尾数≥6时进位。当尾数恰为5时，则应视保留的末位数是奇数还是偶数，5前为偶数应将5舍去，5前为奇数则将5进位。

八、实验报告的书写及格式

1. 实验报告的书写格式 正确书写实验报告是实验教学的主要内容之一，也是基本技能训练的需要。因此，完成实验报告的过程，不仅仅是学习能力、书写能力、灵活运用知识能力的培养过程，而且也是培养基础科研能力的过程。因此，必须完整准确、严肃认真地如实填写。一份完善的实验报告应包括以下6个部分。

（1）实验目的　简述实验的目的要求。

（2）实验原理　简明扼要地说明实验有关的基本原理、性质、主要反应式及方法等。

（3）实验内容　对于实验现象记录与数据记录，要尽量用简洁的形式表示，比如表格、框图、符号等形式，如5滴简写为"5d"，加试剂用"+"，加热用"Δ"，黄色沉淀用"↓黄"、棕红色气体放出用"↑棕红"表示，试剂名称和浓度则分别用化学符号表示。内容要具体详实，记录要表达准确，数据要完整真实。

（4）解释、计算与结论　对实验记录要做出简要的解释或者说明，要求做到科学严谨、简洁明确，写出主要化学反应、离子反应方程式；数据计算结果可列入表格中。

（5）问题与讨论　主要针对实验中遇到的问题提出自己的见解和收获，定量实验则应分析出现误差的原因，对实验的方法、内容等提出改进意见。

（6）完成实验思考题。

2. 示例 实验报告的书写示例如下。

制备型实验报告

一、实验目的（简述实验的目的要求）

二、实验原理（简要说明实验有关的基本原理、主要反应式）

以药用氯化钠的制备反应为例：

$$\text{粗盐}\begin{cases}\text{不溶性杂质，如泥沙等，过滤除去}\\\text{可溶性杂质}\begin{cases}SO_4^{2-}: BaCl_2\\Fe^{3+}, Ca^{2+}, Mg^{2+}: \text{混合碱（}NaOH, Na_2CO_3\text{）}\\\text{重金属离子：}Na_2S\\K^+, Br^-, I^-: \text{留在母液过滤除去}\end{cases}\end{cases}$$

三、实验仪器与材料（简述实验所需的试剂、药材及仪器规格）

四、实验步骤（简单流程图）

30g粗盐 $\xrightarrow[100ml]{+H_2O}$ 溶解/加热 $\xrightarrow[3\sim5ml]{+25\%BaCl_2}$ 加热近沸 → 静置分层

→ 检验沉淀是否完全 $\xrightarrow[pH=11]{+\text{混合碱}}$ 加热至沸 $\xrightarrow[1d]{+Na_2S}$ 静置冷却

过滤/倾滤法 → 得到滤液 $\xrightarrow[pH=4]{+HCl}$ 加热蒸发至糊状 $\xrightarrow{\text{减压过滤}/\text{趁热}}$ 得到半干燥晶体

于蒸发皿中/炒干 → 冷却 → 称重 → 计算产率

五、实验现象与记录

1. 得到的产物颜色、形态、产物的质量、产率

2. 实验总结：

六、实验思考题

定性检查型实验报告

一、实验目的（简述实验的目的要求）

二、实验原理（简要说明实验有关的基本原理、主要反应式）

三、实验仪器与材料（简述实验所需的试剂、药材及仪器规格）

四、实验步骤（按表格形式罗列）

以氧化还原反应实验为例：

1. 电极电势和氧化还原反应

实验步骤	实验现象	解释及反应方程式
(1) $KI+FeCl_3+CCl_4+H_2O$ $KBr+FeCl_3+CCl_4+H_2O$	CCl_4层为紫红色 CCl_4层为无色	$2Fe^{3+}+2I^- \rightleftharpoons 2Fe^{2+}+I_2$ —
(2) $FeSO_4+$碘水$+CCl_4$ $FeSO_4+$溴水$+CCl_4$	CCl_4层为紫红色 CCl_4层为无色	— $2Fe^{2+}+Br_2 \rightleftharpoons 2Fe^{3+}+2Br^-$

结论：

2. 实验记录

实验步骤	实验现象	解释及反应方程式

结论：

五、实验思考题

定量测定型实验报告

一、实验目的（简述实验的目的要求）

二、实验原理（简要说明实验有关的基本原理、主要反应式和测定方法）

三、实验仪器与材料（简述实验所需的试剂、药材及仪器规格）

四、实验记录和处理

以醋酸电离度和电离平衡常数测定实验为例：

醋酸溶液的 pH 测定及其标准平衡常数、电离度的计算表　　$t = \_\_\_\_$ ℃

滴定序号	1	2	3	4
HAc 的稀释倍数	$c/500$	$c/200$	$c/100$	$c/50$
$c(HAc) / mol \cdot L^{-1}$				
pH				
$c(H^+) / mol \cdot L^{-1}$				
$\alpha/\%$				
$K_a^{\ominus}$				
$\overline{K_a^{\ominus}}$				

实验总结：

五、实验思考题

Chapter 1　Introduction

1. Objectives and Requirements of Inorganic Chemistry Experiment Teaching

Chemistry is a natural science based on experiments. Despite the rapid development of modern science and technology, modern chemistry has entered the stage of paying equal attention to theory and practice. Experiment has its irreplaceable role, and it is always the only standard to test whether the theory is right or not. Chemical education in chemical experiment is one of the most effective forms. Under the guidance of teachers, and through the experiment operation, observation, recording data, processing data, and writing test report, the students can use their brains to solve all kinds of problems, and their intelligence factors can be developed fully.

Inorganic chemistry is the first basic chemistry course for students majoring in pharmacy. In order to master the basic knowledge and theory of inorganic chemistry, inorganic chemistry experiment is essential and important. In inorganic chemistry experiments, teachers should give full play to their guiding role and mobilize students' initiative. Both teaching and learning should work hard and pay enough attention to the experimental course.

1.1 Objectives of Inorganic Chemistry Experiment Teaching

(1) Students can acquire a large amount of perceptual knowledge in inorganic chemistry experiments. By induction and summary, they can rise from perceptual knowledge to rational knowledge, and have a deeper understanding of the basic theories and basic knowledge. Through experiments, theory and practice are combined to cultivate students' ability to combine theory with practice, as well as their ability to analyze and solve problems.

(2) Through the training of basic operation skills of inorganic chemistry experiments, students can master the basic operation methods and skills of inorganic chemistry experiments, master the important properties and reaction rules of common elements and their compounds, and understand the general purification and preparation methods of inorganic compounds, laying a solid foundation for the subsequent experiments of various courses.

(3) To train the students' ability to experiment independently, observe and record experimental phenomena, process experimental data correctly and write experimental reports.

(4) To train the students' strict concept of "quality" and "quantity", so that students establish a rigorous scientific attitude.

(5) To develop the good habit of conducting scientific experiments accurately and carefully.

(6) Students can receive standardized and systematic chemical experiment training. To train gradually the students' scientific logical thinking method, scientific attitude of being rigorous and earnest, seeking truth from facts and being meticulous, exploration and cooperation, and the ability to think and practice innovatively.

Experiment is an important part to improve students' comprehensive ability. It not only imparts knowledge, but also trains operation skills. Therefore, inorganic chemistry experiment course is of great significance in the subsequent study of professional experiment courses.

1.2 Requirements of Inorganic Chemistry Experiment Teaching

(1) Teachers should be serious, responsible and strict with students. In particular, they should pay attention to the training of students' basic operation skills of their experimental work ability and good work habits, and run through each specific experiment.

(2) Teachers should insist on asking questions about the preview before each experiment and explaining the key content; in the process of experiment, teachers should conduct careful, and answer questions and solve problems as soon as possible. After experiment, teachers should summarize and discuss, optimize the teaching method, and carefully correct the experiment report.

(3) Students should make it clear that the training of basic operating skills and experimental ability of junior students are the necessary foundation for senior experimental courses and even for scientific research and practical work in the future.

(4) Students should be strict with themselves for all aspects of the experiment. For each experiment, students should not only understand and master the principles and methods of the experiment, but also carry out serious training on the basic operation and meticulous practice on each operation.

(5) Students should develop good work habits and work style during the experiment, such as clean and tidy experiment table, orderly storage of instruments, no litter, fixed-point recovery of experimental reagents, etc.

(6) Pay attention to all aspects of the experiment, especially the safety operation standard of the laboratory must be familiar with.

2. Study Method of Inorganic Chemistry Experiment

To achieve the above learning objectives, we should not only have a correct learning attitude, but also need to have a correct learning method. To do a good inorganic chemistry experiment, we must master the following three important steps.

Step 1: the preview before the experiment is a very important part to ensure experiment is done well. The following requirements should be met through the preview.

(1) Read the experimental materials and watch the teaching videos of relevant experimental operations carefully, pay attention to the interrelation between the experimental materials and theoretical materials, and use the theories to guide the experiment.

(2) Define the purpose of the experiment.

(3) Be familiar with the experimental content, basic principles, experimental steps, basic operations and precautions.

(4) Think carefully about the questions in the experimental textbook, and make their own inferences and explanations about the possible experimental phenomena. For the problems that they can't solve, they can also get the answers through group discussion.

(5) Write an experimental report.

Step 2: during the experiment, students should abide the laboratory rules and accept the guidance of teachers. They should strictly operate according to the methods, steps and reagent dosage specified in the experimental teaching materials and standardized teaching videos. They should do the following:

(1) Operate carefully, observe carefully, carry out in strict accordance with the basic operating

procedures of each experiment, and record the phenomena and experimental data observed in the experiment truthfully.

(2) If the experimental phenomena or results are not consistent with the expectations, the reasons should be analyzed carefully, the experimental errors should be corrected in time, and whether the experiment should be redone should be considered.

(3) Report any difficult problems or emergencies to the teacher in time.

(4) Keep silence during the experiment and work rules of the laboratory.

Step 3: after experiment, the experimental conclusion should be given according to the experimental phenomenon, or the experimental data should be processed and calculated. The experiment report should be completed independently, and the report should be given to the instructor for evaluation.

(1) If the experiment fails, the reasons for the failure should be summarized in time to find out the operational errors and shortcomings in the experiment, so as to avoid the recurrence of similar problems.

(2) If there are experimental phenomena, explanations, conclusions, data, etc. that do not meet the requirements, we should consider redoing the experiment; if there are plagiarized experimental reports, tampered with experimental data, scribbled people rewrite the report.

(3) When writing the experimental report, the handwriting should be correct, concise, neat and clean.

3. Laboratory Rules

(1) The laboratory is a place for teaching and scientific research and is generally not used for other purposes.

(2) All students in the laboratory must abide by the discipline, keep quiet, concentrate their thoughts, operate carefully, observe carefully, think actively and record truthfully.

(3) Before the experiment, students must preview the experiment content, clarify the purpose and requirements, be familiar with the method steps, and master the basic principles.

(4) Use the experimental instruments and equipments correctlys. When using precision instruments, it is necessary to operate them in strict accordance with the operating procedures and be careful to avoid damaging the instruments due to carelessness. If the instrument is found to be faulty, it shall be stopped immediately and reported to the instructor for troubleshooting in time.

(5) Chemicals should be taken in prescribed amounts, and the reagents cannot be put back into the original reagent bottle to avoid contaminating the reagents. The equipment used to take chemicals should be kept clean and dry to ensure that the reagents are not contaminated and the concentration is constant. The stopper should be closed immediately after taking the chemicals to avoid misplace the stopper and contaminate the medicine.

(6) Before the experiment, check whether the required instruments are complete and damaged, so that they can be filled and replaced in time. During the experiment, the experimental table should be kept clean and tidy. After the experiment, the instruments and chemicals should be put back in place.

(7) Waste solids, paper, glass slag, matchsticks, etc. should be poured into the waste basket; the waste liquid should be poured into the designated waste liquid recycling barrel, and should not be poured into the sewer, and the highly toxic waste liquid should be uniformly treated by the laboratory.

(8) After the experiment, keep the laboratory clean, check the safety of water, electricity and gas, and close the doors and windows.

(9) All items (instruments, chemicals, products, etc.) in the laboratory should not be taken out of the

laboratory. Before leaving the laboratory, the instructor's signature is required.

(10) In case of an accident, keep calm and don't panic; report to the teacher in time for first aid and treatment in case of occasional burns, scalds and cuts.

4. Laboratory Safety Rules

Some of the chemicals used in inorganic chemistry experiments are flammable and explosive, and some are corrosive and toxic. Therefore, special attention must be paid to safety in the experiment, and "safety" is very important!

(1) Before entering the laboratory, we should check the storage location of the fire-fighting facilities and be familiar with the emergency access, and learn how to use the fire-fighting equipment.

(2) Potassium, sodium and white phosphorus are easy to burn when exposed to air, so potassium and sodium should be stored in kerosene, while white phosphorus can be stored in water. Use tweezers when taking them. Some organic solvents (such as ether, ethanol, acetone, benzene, etc.) are extremely flammable. When using, they must be kept away from open fire and heat source. After using, close the bottle plug immediately and store in a safe place.

(3) Gas containing oxygen is easy to explode in case of fire, and it must not be close to the fire during operation. Before igniting hydrogen, the purity must be checked and ensured. Silver ammonia solution can't be retained, because it will become explosive substance after long time. Some strong oxidants (such as potassium chlorate, potassium nitrate, potassium permanganate, etc.) or their mixtures cannot be ground, otherwise they will cause explosion.

(4) Equipped with necessary goggles. When pouring chemicals or heating liquid, it is easy to splash out, do not look down on the container. Especially concentrated acids and alkalis are highly corrosive. Do not splash them on skin or clothing. Pay attention to eye protection. When diluting them (especially concentrated sulfuric acid), they should be slowly poured into the water instead of the opposite to avoid splashing. When heating the tube, remember not to make the tube mouth toward yourself or others.

(5) Experiments that produce pungent and toxic gases should be conducted in a fume hood. When smelling the smell of gas, you can only gently stir up the air with your hands to let a small amount of gas enter your nostrils.

(6) Toxic reagents such as arsenic, chromium, cyanide, mercury and their compounds should be strictly prevented from entering the mouth or wounds, and the waste liquid should not be discharged into the sewer.

(7) Metallic mercury is volatile and easy to enter the human body through the respiratory tract, and gradually accumulates to cause chronic poisoning, so the experiment of metal mercury is particularly careful not to spill metal mercury on the table or on the ground. Once spilled, it must be collected as much as possible and covered with sulfur powder in the spilled place to convert metallic mercury into non-volatile and insoluble mercury sulfide.

(8) Use specified chemical spoons or containers to take chemicals. Never taste chemicals. It is not allowed to mix various chemicals casually to avoid accidents. The self-designed experiment can only be carried out after obtaining the teacher's permission.

(9) Do not touch the electrical plugs with wet hand. The human body should not touch electrical objects. After the experiment, the power supply should be cut off.

(10) It is strictly forbidden to eat, drink or smoke in the laboratory. Do not bring food or tableware

into the laboratory. Wash your hands after the experiment.

5. Deal with Laboratory Accidents

There are a large number of inflammable, explosive, toxic and harmful chemicals in the chemical laboratory, so it is necessary to do a good job in fire prevention, explosion prevention and electric shock prevention during chemical experiments.

5.1 Fire Prevention

(1) Keep away from the fire source when doing experiment, do not put flammable solvent in a wide-mouth container for direct fire heating; heating must be done in a water bath and should not be ignited when there is an exposed flammable solvent nearby.

(2) When conducting experiments on inflammable substances, it is necessary to form the habit of first removing flammable substances such as alcohol.

(3) When distilling inflammable organic matter, the device should not leak air. If the leakage occurs, the heating should be stopped immediately and the cause should be checked.

(4) When a large amount of flammable liquid is used, it should be carried out in a fume hood or designated place, and there should be no fire source in the room.

(5) Do not throw away lighted matchsticks or pieces of paper with sparks on them.

(6) The experimenter is not allowed to leave the laboratory without permission when the fire is heated.

5.2 Explosion-proof

(1) Flammable and explosive solvents (such as ether, gasoline, etc.) should not be close to the ignition source.

(2) Do not stress or impact explosive solids (such as heavy metals, acetylides, picric metal salts, trinitrotoluene, etc.).

5.3 Prevent Electric Shock

(1) Before using the electrical appliances, understand the requirements and matching of the electrical appliances to the power supply, and select the appropriate socket or wire.

(2) The circuit must be checked and plugged in before use. At the end of the experiment, the circuit must be cut off before the wires are removed.

5.4 Treatments

If an accident happens in the course of the experiment, please do not panic. You should be calm and deal with it quickly.

(1) Scald: for minor scald, rinse with water first, and then apply the scald ointment. If the scald is serious, you should go to the clinic for treatment immediately.

(2) Cut: do not touch the wound or wash it with water. If it is a glass wound, the broken glass should be picked out from the wound. Apply purple or red liquid medicine to the wound, sprinkle some anti-inflammatory powder or apply some anti-inflammatory cream if necessary, and then bandage with gauze.

(3) Eye injury: to prevent eyes from being fumigated by irritating gas, and to prevent any chemical reagent, especially strong acid, strong alkali, glass chips and other foreign bodies from entering the eye. Goggles should be worn for special experiments. Once any chemical is spilled into the eye, rinse the victim thoroughly with plenty of water for 15 minutes and take him to hospital. When glass chips enter the eyes, try to keep calm, never rub them with hands, or try not to let others take out the chips, let them

cry, wrap the injured's eyes with gauze, and then send them to the hospital for treatment.

(4) Acid corrosion: on the skin, wash with plenty of water, wash with 5% sodium bicarbonate solution, and then apply ointment; on the eyes, wipe off the acid splashed on the outside of the eyes, rinse with water immediately, use with the eye washer aimed at the eyes, then wash with dilute sodium bicarbonate, and drop a small amount of sesame oil finally; on the clothes, wash with water first, then with dilute alkaline water, and finally with water again.

(5) Alkali corrosion: on the skin, wash with saturated boric acid solution or 1% acetic acid solution, and then apply the ointment; around the eyes, wipe off the alkali outside the eyes, wash with water, neutralize the excess alkali with dilute acid, and then rinse with water.

(6) Fire treatment: first cut off the power supply, turn off the gas, and remove inflammable goods. Call the firemen. When a liquid soluble in water catches fire, a wet cloth or water can be used to extinguish the fire. When the insoluble organic reagent with a density less than water catches fire, use sand to extinguish the fire (do not use water). If the fire is large, use CCl_4 extinguisher or CO_2 foam extinguisher, but do not put it out with water, because the water energy and some chemicals (such as sodium metal) have a violent reaction and cause a bigger fire. CCl_4 fire extinguisher must be used in case of electrical equipment fire. Never use water or CO_2 foam fire extinguisher.

(7) Electric shock: in case of electric shock, the power should be cut off and artificial respiration should be performed when necessary.

(8) Serious injuries: send to the hospital immediately.

6. Treatment of "Three Wastes" in the Laboratory

Chemical experiments often produce toxic gases, liquids and solids, which need to be dealt with in a timely manner. In particular, some highly toxic substances, if discharged directly, may pollute the surrounding air and water and harm human health. Therefore, the waste liquid, waste gas and waste residue can only be discharged after a certain scientific treatment.

6.1 Waste Gas Treatment

Experiments to produce poisonous gases should be carried out in a fume hood. Small amounts of toxic gas can be vented out of the house through exhaust devices and diluted by air. When the amount of toxic gas is large, the experiment must be equipped with absorption or treatment device and then discharged. Such as nitrogen dioxide, sulfur dioxide, chlorine gas, hydrogen sulfide, hydrogen fluoride and so on can be used to catheter into the lye, so that most of its absorption back out, carbon monoxide can be ignited into carbon dioxide.

6.2 Disposal of Waste Liquid

(1) A large amount of waste liquid in chemical experiments is usually waste acid. The waste acid can be filtered by acid-resistant plastic mesh or glass fiber, and the filtrate can be neutralized by adding alkali. After pH is adjusted to 6~8, it can be discharged. A small amount of filter residue can be buried underground.

(2) Waste chromic mixture lotion can be regenerated and reused by potassium permanganate oxidation. Oxidation method: firstly, it was continuously stirred, heated and concentrated at 110~130°C. After water was removed, it was cooled to room temperature, and potassium permanganate powder was slowly added. Add about 10g/1000ml and stir until the solution is dark brown or purplish. Do not overdo it. And then we heat it directly until we have sulfur trioxide present, and we stop heating it. Slightly cold,

through the glass sand core funnel filter, remove precipitation; after being cooled, red chromium trioxide precipitate, and then add an appropriate amount of sulfuric acid to dissolve it, so it can be used.

(3) Cyanide is a highly toxic substance, and waste liquid containing cyanide must be carefully disposed of. The waste solution containing cyanide was adjusted to pH 10 with sodium hydroxide solution, and then 3% potassium permanganate was added to decompose CN^- oxidation. The waste liquid with high content of CN^- was treated by alkaline chlorination, that is, sodium hypochlorite was added above pH 10 to oxidize and decompose CN^-.

(4) The waste liquid containing mercury salt was firstly adjusted to pH 8 ~ 10, and excess sodium sulfide was added to generate mercury sulfide precipitation. Then, ferrous sulfate was added as co-precipitant. The resulting iron sulfide adsorbed and co-precipitated the suspended mercury sulfide particles in the water, discharged the liquid, and the residue was recovered by roasting or made into mercury salt. The mercury content of the clear liquid can be discharged below 0.02mg/L. A small amount of residue can be buried underground, and a large amount can be recovered by roasting, but it must be done in a fume hood.

(5) Calcium oxide was added to the arsenic-containing waste solution and pH was adjusted to 8 to produce calcium arsenate and calcium arsenite precipitation. Or adjust pH>10, add sodium sulfide and react with arsenic, produce refractory, low toxic sulfide precipitation.

(6) Waste liquid containing lead and cadmium was adjusted to pH 8 ~ 10 with lime to precipitate $Pb(OH)_2$ and $Cd(OH)_2$ from Pb^{2+} and Cd^{2+}. Ferrous sulfate was added as co-precipitant.

(7) For waste liquid containing heavy metal ions, the most effective and economical treatment method is to deposit the heavy metal ions into insoluble hydroxide or sulfide by adding alkali or sodium sulfide, and then filter and separate them. A small amount of residue can be buried underground.

(8) Phenol was oxidized to carbon dioxide and water by adding sodium hypochlorite or bleaching powder to waste solution containing phenol at low concentration. High concentration phenol waste water was extracted with butyl acetate and recovered by heavy distillation.

(9) Organic solvent

The combustible organic waste liquid can be completely burnt with oxygen in the combustion furnace. Some waste organic solvents can be treated by recycling.

The waste ether solution is placed in the separation funnel, washed once with water, neutralized, washed with 0.5% potassium permanganate until the purple color does not fade, washed again with water, washed with 0.5%~1% ferrous ammonium sulfate solution, removed the peroxide, washed again with water, dried with calcium chloride, filtered, fractionated, collected the fraction of 33.5~34.5°C.

The ethyl acetate waste solution is washed with water for several times, then diluted with sodium thiosulfate solution for several times to make it fade, then washed with water for several times, distilled, dehydrated with anhydrous potassium carbonate, left for a few days, distilled after filtration, and collected the fraction of 76~77°C.

Chloroform, ethanol, carbon tetrachloride and other waste solutions can be washed with waste liquid and treated with reagents. Finally, the distillates at boiling points can be collected by distillation to obtain reusable solvents. Methods can be found in the relevant data.

6.3 Waste Disposal

The solid waste in the experiment should not be put randomly in order to avoid accidents. The dangerous waste that can give off poisonous gas or spontaneously ignite cannot be thrown into the waste bin and discharged into the waste water pipeline. Waste chemicals that do not dissolve in water mustn't be thrown into waste water pipes. They must be burned in appropriate places or treated chemically to make them harmless. Broken glass and other sharp and angular wastes should not be thrown into waste paper baskets, but should be collected and disposed of in special waste bins.

7. Treatment of Data

7.1 Experimental Data

(1) The experimental record book should be numbered page by page. Students must treat the original records strictly and form the habit of keeping the original records.

(2) The experimental data should be standardized. If there are errors in the recorded data, they cannot be completely erased or covered up. The traces of data changes should be seen so as to analyze the causes of the errors.

(3) The original record of the experiment should be recorded on the left side of the record book completely and truthfully, and the processing data should be written on the right side. It is forbidden to record it on paper and then copy it to the record book.

7.2 Statistics and Arrangement of Experimental Data

(1) Data tabulation

a. Determine the data to be recorded: the physical quantity measured.

b. Pay attention to the significant figures and units of data and necessary notes.

c. The designed table should be convenient for data search and comparison, data calculation and further processing.

(2) Data visualization

Visualization is a simplified form of processing chemical experimental results with straight lines or curves. It is applicable to a situation in which the change of one quantity causes the change of another. The most important feature of visualization is simplicity and intuition.

7.3 Experimental Data Screening and Processing

The general method and idea of screening the experimental data are "five views" : Firstly, it is to check whether the data conforms to the accuracy of the measuring instrument. For example, the accuracy of the mass measured by the pallet balance is 0.1g. If the precision value exceeds this range, the obtained data will be invalid. Secondly, we should see whether the data is within the allowable range of error. If the data obviously exceeds the allowable range of error, it should be discarded. Thirdly, whether the reaction is complete or not, whether the data obtained under the action of insufficient reactant, only the data obtained under the complete reaction can be applied; Fourthly, whether the measurement environment of the data obtained is consistent, especially the gas volume. It can be compared and calculated only when the temperature and pressure are consistent. Lastly, whether the data measurement process is standard, reasonable, error and violation of the measurement rules of the data should be discarded.

7.4 Significant Figures

In chemical experiments, chemical calculations are often carried out according to the data measured in the experiments. In order to obtain accurate results, not only accurate measurements should be made, but also correct records and calculations should be made. Correct recording refers to the number of figures correctly recorded, because the number of figures of the data not only indicates the result, but also reflects the accuracy of the measurement.

The following principles should be followed when determining significant digits:

(1) When recording measurement data, only one estimated figure is allowed at the end of the measured value.

(2) When changing units, the number of significant figures must remain the same. For example: 0.0087g should be written as 8.7mg, 23.6L should be written as 2.36×10^4ml.

(3) For large or small numbers, they can be expressed in exponential form. For example: 0.00025g can be written as 2.5×10^{-4}g. 2500g can be written as 2.50×10^3g when three significant figures are needed.

(4) For pH and pK_a, the significant number only depends on the number of digits in the decimal part, while the value in the integer part only represents the power of the original value. For example, when pH = 10.14, the corresponding $[H^+] = 7.2 \times 10^{-11}$mol/L, and the number of significant figures is 2 instead of 4.

In the process of data processing, the following aspects should be paid attention.

(1) If the measurement result is obtained by adding or subtracting several measurements, the number of significant digits retained depends on the one with the least number of decimal places. In term of multiplication and division, the number of significant figures in the answer is the same as the number of significant figures in the measurement that contains the fewest significant figures.

(2) Some times, round off the figures first, and then calculate. When the number is smaller than 4, delete it; when the number is larger than 6, the preceding number should plus 1. When the number is exactly 5, the reserved last digit should be regarded as odd or even. If the number before 5 is even, the 5 should be omitted, and if the number before 5 is odd, the preceding number should plus 1.

8. Writing and Format of the Experiment Report

8.1 Writing and Format of the Experiment Report

Writing experiment reports correctly is one of the main contents of experiment teaching and is also the need for basic skills training. Therefore, the process of completing the experiment report is not only a training process of learning ability, writing ability, and flexible use of knowledge ability, but also a process of cultivating basic scientific research ability. So, it must be filled in accurately and seriously. A good experiment report should include the following 6 parts.

(1) The purpose of the experiment: briefly describe the purpose of the experiment.

(2) Experimental principle: briefly explain the basic principles, properties, main reaction formulas and methods of the experiment.

(3) Experimental procedure: for the experimental phenomena and data records, try to use simple forms, such as tables, block diagrams, symbols, etc., such as 5 drops abbreviated as "5d", plus "+" for adding reagents, and "∆" for heating, the yellow precipitate is represented by "↓ yellow", the reddish brown gas is emitted by "↑ brown red", and the reagent name and concentration are respectively represented by chemical symbols. The content should be specific and detailed, the records should be accurate, and the data should be complete and true.

(4) Interpretation, calculation and conclusion: a brief explanation should be made on the experimental record, which requires scientific, conciseness and clarity, and the main chemical reaction and ion reaction equations should be written; the data calculation results can be included in the table.

(5) Problems and discussion: mainly provide your own insights and gains for the problems encountered in the experiment. For quantitative experiments, you should analyze the reasons for the errors and put forward suggestions for improving the method and content of the experiment.

(6) Complete questions of the experiment.

8.2 The example of Experiment Report

Preparatory Experiment Report

1. The purpose of the experiment (briefly describe the purpose of the experiment)
2. The experimental principle (briefly explain the basic principles of the experiment, the main reaction formula)

Take the preparation reaction of medicinal sodium chloride as an example:

$$\text{coarse salt} \begin{cases} \text{insoluble impurities, such as sediment, removed by filtration} \\ \text{soluble impurities} \begin{cases} SO_4^{2-}: BaCl_2 \\ Fe^{3+}, Ca^{2+}, Mg^{2+}: \text{mixed alkali (NaOH, Na}_2CO_3) \\ \text{heavy metal ion: Na}_2S \\ K^+, Br^-, I^-: \text{left in the mother liquor for filtration and removal} \end{cases} \end{cases}$$

3. Experimental instruments and materials (briefly describe the reagents, medicinal materials and instrument specifications required for the experiment)

4. Experimental steps (simple flow chart)

$$30\text{g coarse salt} \xrightarrow[100\text{ml}]{+H_2O} \xrightarrow[\text{heating}]{\text{dissolve}} \xrightarrow[3\sim5\text{ml}]{+25\%BaCl_2} \xrightarrow[\text{near boiling}]{\text{heating}} \text{static layering}$$

$$\longrightarrow \text{check whether the precipitation is complete} \xrightarrow[pH=11]{+\text{mixed alkali}} \xrightarrow[\text{to boiling}]{\text{heating}} \xrightarrow[1d]{+Na_2S} \text{static cooling}$$

$$\xrightarrow[\text{decantation}]{\text{filter}} \text{get filtrate} \xrightarrow[pH=4]{+HCl} \xrightarrow[\text{to paste}]{\text{heat and evaporate}} \xrightarrow[\text{while it's hot}]{\text{filter under reduced pressure}}$$

$$\text{semi-dry crystals} \xrightarrow{\text{super dry in evaporating dish}} \xrightarrow{\text{cool down}} \xrightarrow{\text{weighing}} \text{calculate yield}$$

5. Experimental phenomena and records
(1) The color, form, quality and yield of the product obtained
(2) Experimental summary
6. Experimental thinking questions

Qualitative Inspection Experiment Feport

1. The purpose of the experiment (briefly describe the purpose of the experiment)

2. The experimental principle (briefly explain the basic principles of the experiment, the main reaction formula)

3. Experimental instruments and materials (briefly describe the reagents, medicinal materials and instrument specifications required for the experiment)

4. Experimental steps (listed in table format)

Take the redox experiment as an example:
(1) Electrode potential and redox reaction

experimental steps	experimental phenomena	explanation and reaction equation
(1) KI+FeCl$_3$+CCl$_4$+H$_2$O KBr+FeCl$_3$+CCl$_4$+H$_2$O	CCl$_4$ layer is fuchsia CCl$_4$ layer is colorless	$2Fe^{3+} +2I^- \rightleftharpoons 2Fe^{2+}+I_2$ /
(2) FeSO$_4$+ iodine water +CCl$_4$ FeSO$_4$+ bromine water +CCl$_4$	CCl$_4$ layer is fuchsia CCl$_4$ layer is colorless	/ $2Fe^{2+}+Br_2 \rightleftharpoons 2Fe^{3+}+2Br^-$

Conclusion:

(2) Experimental records

experimental steps	experimental phenomena	explanation and reaction equation

Conclusion:

5. Experimental thinking questions

Quantitative Determination Experiment Report

1. The purpose of the experiment (briefly describe the purpose of the experiment)

2. The experimental principle (briefly explain the basic principles of the experiment, the main reaction formula and the determination method)

3. Experimental instruments and materials (briefly describe the reagents, medicinal materials and instrument specifications required for the experiment)

4. Experimental records and processing
Taking acetic acid ionization degree and ionization equilibrium constant determination experiment as an example:

Determination of pH value of acetic acid solution and calculation table of standard equilibrium constant and ionization degree $t = \underline{\quad}$ °C

titration serial number	1	2	3	4
HAc dilution factor	$c/500$	$c/200$	$c/100$	$c/50$
$c(\text{HAc})/\,\text{mol}\cdot\text{L}^{-1}$				
pH				
$c(\text{H}^+)/\,\text{mol}\cdot\text{L}^{-1}$				
$\alpha/\%$				
$K_a^{\ominus}$				
$\overline{K_a^{\ominus}}$				

Experimental summary:

5. Experimental thinking questions

第二章　常用仪器与基本操作

一、称量

（一）直接法

天平零点调定后，将被称物直接放在托盘上（注意化学试剂不能直接接触托盘），按照天平使用操作规范进行称量，所得读数即被称物的质量。这种称量方法适用于称量洁净干燥的器皿、棒状或块状的金属及其他整块的不易潮解或不易升华的固体样品。注意，禁止用手直接取放被称物，可借助汗布手套、垫纸条、用镊子或钳子等适宜的办法。

（二）递减称量法

这种方法称出样品的重量不要求固定的数值，只需在要求的范围内即可，适于称取易吸水、易氧化或易与 CO_2 反应的物质。将此类物质盛在带盖的称量瓶中进行称量，可防止吸潮和防尘。称量顺序如下。

先在称量瓶中装适量试样（如果试样曾经烘干，应放在干燥器中冷却到室温），用洁净的小纸条或塑料薄膜套，套在称量瓶上，再将称量瓶放在分析天平上精确称出其重量，设为 $W_1(g)$。将称量瓶取出，用洁净的小纸条或塑料薄膜套，套在称量瓶盖上，在容器口上方，取下称量瓶盖，用称量瓶盖轻轻地敲瓶的上部，使试样慢慢落入容器中，如图 2-1 所示，然后慢慢地将瓶竖起，用瓶盖敲瓶口上部，使粘在瓶中的试样落入瓶中，盖好瓶盖。再将称量瓶放回天平盘上称量，如此重复操作，直到倾出的试样重量达到要求为止。设倒出第一份试样后称量瓶与试样重为 $W_2(g)$，则第一份试样重 $W_1-W_2(g)$（操作如图 2-1 所示）。

图 2-1　从称量瓶中敲出试样示意图

二、无机化学实验常用仪器

仪器	规格	一般用途	使用注意事项
试管 试管架	试管：分硬质、软质；有刻度、无刻度；具支管、无支管等 无刻度试管一般用管口外径×管长（mm）计，如 10mm×75mm、15mm×150mm 等 有刻度试管按容量（ml）计，如 5、10、15ml 等 试管架：木质、塑料或金属	试管：①少量试剂的反应容器，便于操作、观察；②具支试管可以装配气体发生器、洗气装置和检验气体产物 试管架：盛放试管	试管：①硬质试管可以加热至高温，但不宜骤冷，软质试管在温度急剧变化时极易破裂；②加热时试管口不要对人，盛放液体不宜超过容量的1/3。试管架、金属试管架应注意防酸碱腐蚀
离心试管	分有刻度和无刻度两种，规格以容量（ml）计，如 5、10ml 等	（1）少量试剂反应器 （2）分离沉淀	不能直接加热，可以水浴加热
烧杯	以容量（ml）计，如 50、100、500ml 等 分硬质、软质，有刻度和无刻度等	（1）反应容器，易混合均匀 （2）配制溶液	（1）加热前将烧杯外壁擦干，加热时下垫石棉网，使受热均匀 （2）反应液体不得超过烧杯容量的2/3，以免外溢
锥形瓶	以容量（ml）计，如 50、100、250ml 等。分硬质、软质，具塞、无塞等，具塞的又称碘量瓶	（1）反应容器，摇荡方便 （2）适用于滴定操作	不能直接加热，加热时下垫石棉网或置于水浴中加热
点滴板	上釉瓷板，分白黑两种	进行点滴反应，观察沉淀生成和颜色变化	（1）不能加热 （2）不能用于含氢氟酸和浓碱溶液的反应

续表

仪器	规格	一般用途	使用注意事项
量筒　量杯	以所能度量的最大容量（ml）计，如10、50、100ml等	量取一定体积的液体	（1）不能作为反应容器，不能加热，不可量热的液体 （2）读数时视线应与液面水平，读取与弯月面最低点相切的刻度
广口瓶　细口瓶	材料：玻璃或塑料 规格：广口、细口；无色、棕色 以容量计，如125、250、500、1000ml等	广口瓶盛放固体试剂，细口瓶盛放液体试剂	（1）不能加热 （2）取用试剂时，瓶盖应倒放在桌上 （3）盛碱性物质要用橡皮塞或塑料瓶 （4）见光易分解的物质用棕色瓶
药匙	以大小计，由瓷、骨、塑料、金属合金等材料制成	用于取用固体试剂	（1）药匙大小的选择，应以盛取试剂后能放进容器口为宜 （2）取用一种药品后，必须洗净擦干才能取用另一种药品
研钵	材料：瓷质、铁、玻璃、玛瑙等 规格：以钵口径计，如60、75、90mm	（1）研磨固体物质 （2）两种或两种以上药品通过研磨，混合均匀	（1）不能作反应容器用 （2）只能研磨，不能敲打，不能烘干 （3）易爆物质只能轻轻压碎，不能研磨
表面皿	以直径（mm）计，如45、65、75、90mm	（1）少量试剂反应器 （2）作蒸发皿、烧杯等容器的盖子	（1）不能用火直接加热 （2）作盖子用时，其直径应比被盖容器略大
漏斗	以口径和漏斗颈长短计，如6cm长颈漏斗、4cm短颈漏斗	用于过滤或倾注液体	不能用火直接加热
布氏漏斗　抽滤瓶	布氏漏斗：瓷质制或玻璃制，以直径（cm）计 抽滤瓶：以容量（ml）计	减压抽滤	（1）滤纸要略小于漏斗内径，又要把底部小孔全部盖住，以免漏液 （2）先抽气，后过滤。停止过滤时，先放气，再关泵 （3）不能直接加热

续表

仪器	规格	一般用途	使用注意事项
蒸发皿	材料：瓷质，也有石英、铂制品 以上口直径（cm）或容量（ml）计。分平底、圆底两种	（1）蒸发浓缩 （2）反应容器 （3）灼烧固体	（1）耐高温，能直接加热 （2）注意不要碰碎，高温时不能骤冷
容量瓶	以容量（ml）计，分量入式（in）和量出式（ex） 颜色分无色、棕色两种	用于配制标准溶液	（1）不能直接加热，不能代替试剂瓶用来存放溶液 （2）棕色容量瓶用于配制见光易分解的药品标准溶液 （3）瓶塞和瓶是配套的，不能混用
移液管 吸量管	以刻度最大的标度（ml）计，如1、2、5、10、20ml等 无分度的为移液管，有分度的为吸量管	用于精密量取一定体积的液体	（1）不能加热烘干 （2）用时先用少量待取液润洗3次 （3）一般移液管残留的最后一滴液体，不要吹出；但刻有"吹"字的完全流出式移液管除外
称量瓶	分高形、扁形，以外径（mm）×瓶高（mm）计	用于准确称取一定量的固体	（1）瓶与盖是配套的，不可混用 （2）不可用火加热 （3）不用时应洗净，在磨口处垫一张小纸条
干燥器	以上口径（mm）计，分普通、真空干燥器两种，颜色有无色、棕色两种	（1）存放需要保持干燥的仪器或试剂 （2）定量分析时，将灼烧过的坩埚放在其中冷却	（1）灼烧过的物品应稍冷，再放入干燥器内 （2）干燥剂要及时更换，干燥器盖子的磨口处应均匀涂抹凡士林 （3）见光易分解的样品宜用棕色干燥器

续表

仪器	规格	一般用途	使用注意事项
坩埚	材料：瓷质、石英、氧化锆、铁、镍、铂等 以容量（ml）计，常用的为30ml	耐高温，用于灼烧固体	（1）可放在泥三角上，直接用火加热 （2）灼热的坩埚用坩埚钳夹取，放置于石棉网上 （3）灼热的坩埚不可骤冷，避免溅上水
坩埚钳	铁或铜合金制品，表面常镀铬、镍	高温时，夹取坩埚或坩埚盖	（1）夹取灼热的坩埚时，需将尖部预热，以免坩埚因局部骤冷而炸裂 （2）放置时应将尖部朝上放置 （3）避免接触腐蚀性液体
三脚架	铁制品，有大小、高低之分	放置较大或较重的加热容器	先放石棉网，再放加热容器，水浴锅除外
泥三角	有大小之分，以每边边长（cm）计	用于盛放加热的坩埚和小蒸发皿	（1）灼热的泥三角避免滴上冷水，以免瓷管破裂 （2）选择泥三角时，要使搁在上面的坩埚所露出的上部不超过本身高度的1/3
铁架、铁圈和铁夹	铁制品，夹口常套有橡胶或塑料 铁架台以高度（cm）计；铁圈以直径（cm）计；铁夹以大小计，有双钳、三钳、四钳之分	（1）固定反应容器用 （2）铁圈可代替漏斗架或泥三角支撑架使用	（1）应先将铁夹等升至合适的高度，并旋紧螺丝，使之牢固后再进行实验 （2）固定仪器在铁架台上时，重心应落在铁架台底盘的中心处
石棉网	由铁丝编成，中间涂有石棉，有面积大小之分，其大小以石棉层的直径计，如10、15cm等	加热玻璃仪器时垫在底部，使其受热均匀	（1）避免石棉网浸水，以免石棉脱落、铁丝锈坏 （2）因石棉致癌，国外已用高温陶瓷代替

续表

仪器	规格	一般用途	使用注意事项
毛刷	以大小和用途计，如试管刷、烧杯刷	用于洗刷常用玻璃仪器	洗涤试管时，要把前部的毛捏住放入试管，以免铁丝顶端将试管底戳破
洗瓶	塑料制品，以容量（ml）计，如 500ml	盛装蒸馏水或去离子水，洗涤沉淀和容器用	（1）不能装自来水用 （2）塑料洗瓶不能加热
试管夹	有木质和金属制品，形状大同小异	用于加热时加持试管	（1）夹在试管上端（距试管口约 2cm 处） （2）从试管底部套上或者取下试管架 （3）加热时手握试管夹的长柄，不要同时握长柄和短柄

注：仪器所用材料无注明者皆为玻璃。所列规格为常用的规格。

三、仪器的洗涤与干燥

（一）玻璃仪器的清洗

化学实验中经常用到玻璃仪器，仪器洁净与否对实验结果的准确性有很大影响。因此，洗涤玻璃仪器是一项必须的化学实验准备工作。针对不同污物性质和沾污程度，可以选择不同的清洗方法。

1. 一般洗涤 实验室常用的烧杯、烧瓶、锥形瓶、量筒、表面皿、试剂瓶等玻璃器皿的清洗，可先把仪器和毛刷淋湿，然后用毛刷蘸取去污粉刷洗仪器的内、外壁，至玻璃表面的污物除去，再用自来水冲洗。如果仪器内壁附着的水既不聚成水滴，也不成股流下，而是均匀分布形成一层水膜，表示仪器已洗干净。

2. 洗液洗涤 移液管、容量瓶、滴定管等具有精密刻度的量器内壁不宜用刷子刷洗，也不宜用强碱性溶剂洗涤，以免损伤量器内壁而影响准确性。通常用含 0.5% 的合成洗涤剂的水溶液浸泡或将其倒入量器中晃动几分钟后弃去，再用自来水冲洗干净。有的污垢难以洗净，可针对其性质选用适当的洗液进行洗涤，用腐蚀性洗液洗涤时不可用毛刷刷洗。如果是酸性的污垢，可用碱性洗液洗涤；反之，碱性的污垢可用酸性洗液除去。氧化性污垢可以用还原性洗液洗涤；

还原性污垢可用氧化性洗液洗涤。

若污物是无机物，一般选用铬酸洗液洗涤。铬酸洗液由浓硫酸和饱和重铬酸钾溶液混合配制而成，它主要用于定量实验中如滴定管、移液管、容量瓶等仪器及形状特殊的仪器的洗涤。若污物是有机物，一般选用 $KMnO_4$ 碱性洗液洗涤。洗液具有强腐蚀性，使用时如果不慎将洗液洒在衣物、皮肤或桌面时，应立即用水冲洗。用后的洗液应倒回原瓶，可反复多次使用，多次使用后，铬酸洗液会变成绿色（Cr^{3+}）；$KMnO_4$ 洗液会变成浅红色或者无色（Mn^{2+}），底部有时会出现 MnO_2 沉淀。这时洗液已不具有强氧化性，不能再继续使用。失效的洗液应处理后倒在废液缸里，不能倒入水槽，以免腐蚀下水道和污染环境。

3. 特殊污垢的洗涤 有些仪器上常常沉积一些已知化学成分的污垢，这时就需要视污垢的性质选用合适的试剂，通过化学反应而除去。例如，AgCl 沉淀，可以选用氨水洗涤，做银镜反应实验时在试管底部沉积的银可用稀硝酸除去。

如果不是十分需要，不宜盲目使用各种化学试剂和有机溶剂来清洗仪器，这样不仅造成浪费而且还可能带来危险。

除了上述的清洗方法外，还有超声波清洗法。将需清洗的仪器放在装有洗涤剂的容器中，接通电源，利用超声波的振动，达到洗涤的目的。

日常实验中应养成玻璃仪器用毕立即清洗的习惯，因为污垢的性质在当时是清楚的，用适当的方法进行洗涤容易办到，若是放久了，将会增加洗涤的难度。

检查玻璃器皿是否洗净的方法是加水倒置，水顺着器皿壁流下，内壁被均匀湿润着一层薄的水膜，且不挂水珠，否则需要再进行洗涤，这样洗净的玻璃仪器可供一般化学实验使用。若是洗涤用于精制或有机分析的器皿，除用上述方法处理外，还必须用蒸馏水或去离子水冲洗，以除去自来水引入的杂质。用蒸馏水洗涤的办法应采用"少量多次"的原则，一般用洗瓶将蒸馏水均匀地喷射在仪器内壁并不断地转动仪器，再将水倒掉。如此重复 2~3 次即可。

（二）玻璃仪器的干燥

有些仪器清洁后即可使用，但有些化学实验所需仪器必须是干燥的。例如，定量分析中的非水滴定，一些有机化学实验所用玻璃仪器等；仪器的干燥与否，有时是实验成败的关键。干燥的方法有以下几种。

1. 晾干 对不急用的仪器，洗净后倒立放置在适当仪器架上，让其在常温下自然干燥。

2. 烘干 把洗净的玻璃仪器放入电热恒温干燥箱内烘干。放入前先将水沥干，无水珠下滴时，将仪器口向上，放入烘箱内，并且是自上而下依次放入，将烘箱温度调节为 105~110℃，烘 1 小时左右。带有磨口玻璃塞的仪器，烘干时必须取出玻璃塞，玻璃仪器上附带的橡胶制品在放入烘箱前也应取下。有时也可将仪器放入托盘内。

当烘箱已工作时，不能往上层放入湿的器皿，以免水滴下落，使热的器皿骤冷而破裂。仪器烘干后要待烘箱内的温度降低后才能取出，取玻璃仪器时，要注意防止烫伤。切不可将很热的玻璃仪器取出直接接触冷水、瓷板等低温台面或冷的金属表面，以免骤冷使之破裂。

3. 吹干 将洗净的玻璃仪器中的水倒尽后放在气流干燥器上用冷、热风吹干或用吹风机把仪器吹干。

4. 有机溶剂干燥 该法适用于仪器洗涤后需要立即干燥使用的情况。将洗净的玻璃仪器中的水尽量沥干，加入少量 95% 的乙醇摇洗并倾出，再用少量丙酮摇洗一次（需要的话最后再用乙醚摇洗），然后用电吹风冷风吹 1~2 分钟（有机溶剂蒸气易燃烧和爆炸，故应先吹冷风），待大部分溶剂挥发后，再用热风吹至完全干燥，最后再用冷风吹去残余蒸气，以免又冷凝在仪器内，并使仪器逐渐冷却。

需要注意的是，带有刻度的计量仪器不可用加热的方法进行干燥，以免影响仪器的精密度。具有挥发性、易燃性、腐蚀性的物质不能放入烘箱。用乙醇、丙酮淋洗过的仪器晾干或者用无尘布擦干即可使用，严禁直接放入烘箱烘干，以免发生着火或爆炸。磨口玻璃仪器和带有活塞的仪器洗净后放置时，应该在磨口处（如容量瓶、酸式滴定管等）垫上小纸片，以防长期放置后瓶塞粘上不易打开。

四、水及化学试剂的分类、管理、取用

（一）化学试剂的分类

化学试剂是一大类精细化学品的总称，是工业、农业、医疗卫生及科研教学等行业必须的药剂。化学试剂的品种极多，从不同角度可以分成很多种类型。根据国家标准《化学试剂分类》（GB/T 37885—2019)的规定，化学试剂按照用途可分成十大类。

（1）基础无机化学试剂。

（2）基础有机化学试剂。

（3）高纯化学试剂。

（4）标准物质/标准样品和对照品（不包括生物化学标准物质/标准样品和对照品）。

（5）化学分析用化学试剂。

（6）仪器分析用化学试剂。

（7）生命科学用化学试剂（包括生物化学标准物质/标准样品和对照品）。

（8）同位素化学试剂。

（9）专用化学试剂。

（10）其他化学试剂。

在实际应用中，具体的化学试剂按照国家标准的不同，根据产品纯度及杂质含量可分为优级纯（G.R）、分析纯(A.R)、化学纯(C.P)、实验试剂(L.R)，其标签颜色分别规定为绿色、红色、蓝色、棕色或黄色。大部分常用试剂有国家规定的等级标准，部分没有国家标准的可遵循行业标准或者企业标准。四种标准的化学试剂的生产成本和销售价格依次呈递减状态，所以在选用试剂时要综合考虑，选用合适的试剂以降低使用成本。

一般在无机化学实验中，化学纯试剂即可满足实验要求。分析纯适用于分析工作，对于要求更高的精密分析则要求使用优级纯试剂。随着科学技术的发展，一些实验对化学试剂提出了更高的要求，因此出现一些专门用途的试剂如高纯试剂、色谱纯试剂等。

从化学试剂性质的角度，可将其分为一般试剂和危险试剂。根据《化学品分类和危险性公示通则》（GB 13690—2009）的规定，危险试剂可分为理化危险、健康危险及环境危险三大类。理化危险主要指爆炸物、易燃物、自燃物、氧化及过氧化物。健康危险性主要指急性毒性、致癌性、生殖毒性、呼吸毒性等。环境危险则主要指危害水生环境、急性水生中毒等。

（二）化学试剂的管理

化学实验室试剂的种类繁多，且多数化学试剂具有危险性。对化学试剂科学、规范的管理，是实验工作顺利展开的保障，也是实验室安全管理工作的重要组成部分。应选择通风、干燥、远离热源和火源的专用试剂柜存放化学试剂，柜内试剂应按其性质分格放置，固体试剂与液体试剂分柜存放。试剂可按照英文字母排列或者按照试剂特点划分类别，并且应按照生产日期、保质期以先进先出的原则入库出库。如条件允许可以采用进销存软件进行统一管理。

1. 一般试剂的存放 一般试剂应贮存在阴凉通风处，远离热源、水源，避免潮湿环境。贮藏位置应尽量避免人员活动，防止药品碰撞跌落。这类试剂主要包括无机盐、难挥发有机物、低燃点有机物等。虽然一般试剂对贮存条件要求较低，但应根据试剂性质采用不同的保存方法。

（1）遇光易变质的试剂　部分试剂遇强光易引起试剂本身分解变质，如银的氧化物和盐类、高汞和亚汞的氧化物、卤化物（除氟外）及其盐类、溴和碘的一些盐类。此类试剂应采用棕色玻璃瓶保存，外用黑纸包裹，并避光置于试剂柜中。

（2）遇热易变质的试剂　多数生物制品在温度较高条件下易失活、分解发霉。部分试剂在常温下易发生反应，如甲基丙烯酸甲酯、苯乙烯、丙烯腈、乙烯基乙炔及其他可聚合的单体，其存放温度应在10℃以下。

（3）易冻结试剂　部分试剂熔点处在室温范围，当气温高于其熔点，或下降到凝固点以下时，则试剂由于熔化或凝固而发生体积的膨胀或收缩，易造成试剂瓶的炸裂。如硫酸、苯酚、冰乙酸等。冰乙酸低温时凝固成冰状，俗称冰醋酸。凝固时体积膨胀可能导致容器破裂。此类试剂应采取防瓶裂措施。

（4）易风化试剂　在空气过于干燥时，含有结晶水的试剂可逐渐失去结晶水而变质，使有效成分含量发生变化。如结晶碳酸钠、结晶硫酸铝、结晶硫酸镁、胆矾、明矾等，此类试剂应密封保存。

（5）易潮解试剂　部分试剂易吸收空气中的水分发生潮解、变质，导致有效成分降低甚至发生霉变等。如氯化铁、无水乙酸钠、甲基橙、琼脂、还原铁粉、铝银粉等。易潮解或受潮后变质的试剂，应贮于干燥器内。

2. 化学危险品的存放

（1）易燃类

易燃类液体：部分试剂易挥发成气体，遇明火燃烧，所以通常把闪点在25℃以下的液体列入易燃类。易燃液体主要是有机溶剂，如甲醇、乙醇、石油醚、乙酸乙酯、乙醚、丙酮、二硫化碳、苯、甲苯等，这些液体应单独存放，并注意保持阴凉、通风，特别要注意远离火源。

易燃类固体：无机物中的硫黄、红磷、镁粉和铝粉等，燃点低，属于易燃类固体。存放处应通风、干燥。白磷在空气中可自燃，因此应保存在水里，放置于避光、阴凉处。定期检查瓶中的水量，防止水分蒸发使白磷露出水面。

（2）剧毒类　剧毒类指由消化道侵入、少量即可引起中毒致死的试剂。如氰化物、三氧化二砷及其他砷化物、氯化汞、硫酸二甲酯、可溶性钡盐、铅盐、锑盐、某些生物碱等。剧毒类试剂应锁在专门固定的毒品柜中，由专人负责保管，并建立使用消耗废液处理制度，皮肤有伤口时禁止使用这类物质。

（3）强腐蚀类　把对皮肤、黏膜、眼、呼吸道和物品等有强腐蚀性的液体和固体归类于强腐蚀性物质。例如强酸、强碱、液溴、三氯化磷、五氧化二磷、无水三氯化铝、氨水、硫化钠、苯酚、水合肼等。这类药品应选用抗腐蚀性的材料来储存并与其他药品隔离放置。

（4）易爆类　金属钾、钠、钙、电石和锌粉等遇水发生剧烈反应，并放出可燃性气体，极易引起爆炸。钾、钠应保存在煤油里，电石、锌粉应放在干燥处。三硝基甲苯、三硝基苯、叠氮或重氮化合物等，要轻拿轻放。这些物质的存放应与易燃物、强氧化剂等隔开。长时间不用时，应将其密封保存，并将盛装这些物质的容器放在用水泥或砖砌成的槽中。

（5）强氧化剂类　强氧化剂包括过氧化物及其盐（过氧化氢、过氧化钠、过氧化钡）、强氧化性的含氧酸及强氧化性的含氧酸盐（硝酸盐、氯酸盐、重铬酸盐、高锰酸盐）。当受热、撞击或混入还原性物质时，就可能引起爆炸。要求于阴凉通风处存放，最高温度不得超过30℃，并且要与酸类及木屑、炭粉、硫化物、碳水化合物类等易燃物、可燃物或易被氧化物等隔离，注意散热。

（6）放射性类　放射性物质应存放在铅器皿中，操作这类物质需要具备特殊防护设备及相关知识，以保证人身安全，同时防止放射性物质的污染和扩散。

（三）化学试剂的取用

化学试剂取用前需明确试剂的名称、纯度、有效期，取用试剂时需避免污染标签。操作时，试剂瓶瓶盖反放在实验台上，避免瓶盖被污染。不得用手碰触化学试剂，取完试剂后，将原瓶盖盖回原试剂瓶，将试剂瓶放归原处。

1. 固体试剂的取用

（1）用干净、干燥的药匙取固体试剂，专匙专用，用过的药匙需要经过清洗、干燥后才能再次使用。

（2）取用试剂时，若超过需要的量，多取的试剂不可放回原瓶，避免污染。

（3）称取精确质量的固体试剂时，把称量纸放置在天平上，清零，取用试剂，称量。具腐蚀性或易潮解试剂可放在玻璃器皿内进行称量。

（4）向试管内加固体试剂时，用药匙或者将取出的药品放在对折的称量纸上，伸进试管2/3处。试管加入块状固体时，应将试管倾斜，使试剂沿着试管内壁缓慢滑入试管底（图2-2）。

图2-2　固体试剂的取用示意图

（5）颗粒较大的固体，可先在干燥的研钵中研磨之后再进行量取。

2. 液体试剂的取用

（1）取用液体试剂时，需用干净干燥的滴管。滴管取用液体药品时，先用滴管反复吸取几次试剂，使试剂与滴管内壁充分接触，避免试剂受热膨胀而滴落。滴管不能伸入所用实验容器中，以免接触容器内壁造成污染（图2-3）。装有药品的滴管不得横置或者管口向上斜放，以免液体滴入到胶头中。试剂瓶的胶头滴管应做到专瓶专用，不得交叉使用。

（2）对于无法用胶头滴管取用试剂的细口瓶，可采用倾注法移取试剂。先将瓶塞取下，反放在桌面上。试剂瓶标签朝向掌心，倾斜试剂瓶，将试剂沿着试管壁倒入试管，或者沿着玻璃棒注入烧杯或者容量瓶（图2-4）。

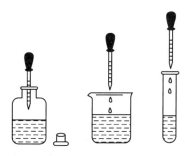

用滴管吸取和滴加试剂（正确）

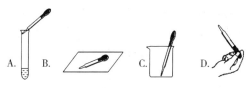

胶头滴管的错误使用

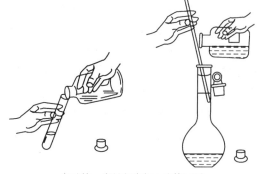

向试管、容量瓶中加入液体试剂

图2-3　胶头滴管使用示意图　　　图2-4　液体取用示意图

（3）在进行性质实验时，对于液体试剂的量取不需要很精确，这时可用滴管近似取用试剂。胶头滴管滴出液滴15~20滴约为1ml。

五、基本度量仪器的使用

（一）量筒

量筒是化学实验室中最常见的度量仪器。量筒有多种规格，使用时应根据需要选择合适量程的量筒。一般来说，量程越大的量筒误差越大。因此在选取量筒时，应让量筒量程尽量接近目标体积，以降低误差。例如需要取用6ml液体时，应选用量程为10ml的量筒。若选取量程为100ml的量筒，则会产生较大误差。读数时，量筒应放置在水平的桌面上，确保视线与量筒内溶液凹液面最低处在同一水平线上，否则会增大测量误差。由于热胀冷缩，温度过高或者过低都会影响测量结果，因此量筒应室温下使用。

（二）滴定管

滴定管是滴定分析中最基本的量器，由长玻璃管身和流速控制开关构成。滴定管身上刻有精密刻度，可以通过滴定前后差值得到消耗溶液的体积。常量分析用的滴定管有50ml及25ml两种规格，它们的最小分度值为0.1ml，加上一位估读其最小计数单位应为0.01ml。此外，还有量程为10ml、5ml、2ml和1ml的半微量和微量滴定管，最小分度值为0.05ml、0.01ml或0.005ml。

根据控制溶液流速的装置不同，滴定管可分为酸式滴定管、碱式滴定管和通用滴定管（图2-5）。装有磨砂玻璃旋塞的为酸式滴定管，可盛放酸性、氧化性溶液及盐类溶液。由于玻璃与碱性溶液可生成硅酸盐将旋塞黏住，碱性溶液应用碱式滴定管盛放。碱式滴定管管身下端连接一乳胶管，乳胶管内有一直径略大于胶管内径的玻璃球用以控制溶液流速。凡可与乳胶管发生反应的氧化型溶液，如$KMnO_4$、I_2等均不能用碱式滴定管盛放。酸碱通用滴定管在结构上与酸式滴定管类似，其旋塞采用聚四氟乙烯制成，具有耐酸、耐碱和耐腐蚀性的特点。同批次滴定管之间的聚四氟乙烯旋塞可以通用，适用于酸性、碱性和强氧化性溶液的滴定。

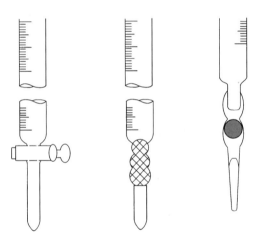

图2-5 滴定管结构示意图

1. 使用前准备

（1）酸式滴定管 使用前应检查玻璃旋塞是否贴合紧密，将旋塞用水润湿后插入旋塞槽内，管中充水至零点。管身用滴定管夹固定。15分钟后漏水不超过1个分度（50ml滴定管为0.1ml）视为合格。为使玻璃旋塞密封和润滑，需要将旋塞用凡士林涂油。首先取出旋塞，用滤纸吸干旋塞和旋塞槽内的水，在旋塞粗的一端和旋塞槽细的一端内壁涂上一薄层凡士林。涂好凡士林的旋塞小心插入旋塞槽内，将旋塞沿着同一方向旋转，直到旋塞部位的油膜均匀透明（图2-6）。如发现旋塞转动不灵活，或者旋塞上出现纹路，表示油涂的不够；若有凡士林从旋塞缝溢出，或者旋塞孔被堵，表示凡士林涂得太多。涂好凡士林后，用乳胶套套在旋塞的末端，以防旋塞脱落破损。将涂好凡士林的滴定管旋塞关闭，充水至零点，并将其固定在滴定管架上，2分钟后无水渗出则将旋塞旋转180°，再次放置2分钟，两次均无渗水则视为合格。若有漏液，则应重新进行涂油工作。

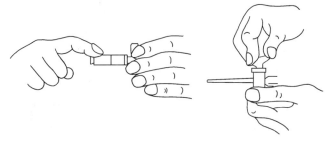

图 2-6 酸式滴定管旋塞抹油示意图

准备完毕的酸式滴定管用自来水冲洗干净，再用铬酸洗液浸泡，倒出洗液后，再次用自来水将残留洗液冲洗干净，然后用蒸馏水冲洗三次。滴定管内壁呈均一水膜，不挂水珠则视为清洗合格。清洗时，先关闭旋塞，装入少量洗涤液，双手持滴定管，边旋转边倾斜滴定管，管身一端稍低，旋塞一端稍高，让洗涤液润湿整个管身。留少许洗涤液，打开旋塞让洗涤液从旋塞端流出，使整个滴定管充分洗涤。

（2）碱式滴定管　选择合适的尖嘴、玻璃珠和乳胶管，合格的碱式滴定管在静止状态下应不漏液，进行滴定操作时应滴液流畅，否则需要重新装配。碱式滴定管清洗方法及要求与酸式滴定管相同。

（3）酸碱通用滴定管　其试漏方法与酸式滴定管相同，如果有漏液现象，可通过调节聚四氟乙烯旋塞的螺母松紧来解决，否则需要更换滴定管。

2. 装液　装入操作溶液前，先将贮液瓶内操作液摇匀，之后将滴定管用操作液润洗 2~3 次，每次约 10ml。之后，由贮液瓶直接将操作液灌入滴定管，（不得借用任何其他器皿，如移液管、烧杯、漏斗，以避免操作液被污染或被稀释），最终使管内液面稍高于最高刻度线。确保装满操作液的滴定管尖嘴内没有气泡，如有气泡必须排除。对于酸式滴定管，右手持管身倾斜 30° 左右，左手迅速打开旋塞，使操作液快速流出，带走气泡。对于碱式滴定管，需将乳胶管向上弯折，挤压玻璃球，使操作液快速冲出，排除气泡（图 2-7）。对于酸碱通用滴定管，排气泡方法与酸式滴定管相同。装液完毕后，调节溶液液面至零刻度线。

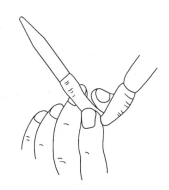

图 2-7 除去碱式滴定管胶管中气泡

3. 读数　将装满溶液的滴定管垂直地夹在滴定管夹上。由于溶液表面张力的作用，滴定管内的液面呈弯月形。无色溶液的弯月面比较清晰，而有色溶液的弯月面清晰程度较差。因此两种情况的读数方法稍有不同。为了正确读数，应遵循以下原则。

读数时滴定管应垂直地面，注入溶液或放出溶液后，需等待 1~2 分钟后才能读数。对于无色溶液或浅色溶液，应读弯月面下缘实线的最低点。为此，读数时，视线应与弯月面下缘实线的最低点在同一水平上。有色溶液，如 $KMnO_4$、I_2 等，视线应与液面两侧的最高点相切。

滴定时，应从零刻度线开始，这样可以固定在某一体积范围内度量滴定时所消耗的标准溶液，减少误差。同时可以避免标准溶液在刻度线范围外被耗尽，造成无法计算滴定前后体积差。

为了协助读数，可采用读数卡。这种方法有利于初学者练习读数。读数卡可用黑纸或用一中间涂有黑长方形的白纸制成。读数时，将读数卡放在滴定管背后，使黑色部分在弯月面下约 1mm 处，此时即可看到弯月面的反射层变成黑色，然后读此黑色弯月面下缘的最低点（图 2-8）。读数应准确到0.01ml。

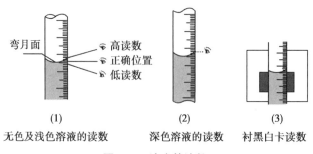

图2-8 滴定管读数

4. 滴定操作 在使用酸式滴定管时应用左手控制滴定管旋塞，大拇指在旋塞靠近身体一侧，食指和中指在旋塞远离身体一侧。手指略微弯曲，轻轻向掌心施力，以免旋塞松动或被顶出。右手握锥形瓶持续向同一方向摇动，注意不能前后振动，否则会溅出溶液。滴定初期滴定速度可稍快，一般为10ml/min，每秒3~4滴。当标准液进入锥形瓶造成局部指示剂变色，振荡后颜色缓慢消失，则说明接近滴定终点。临近滴定终点时，应一滴或者半滴地加入，并用洗瓶吹入少量水冲洗锥形瓶内壁，使附着的溶液全部流下，然后摇动锥形瓶，如此继续滴定直至终点为止（图2-9）。

在使用碱式滴定管时，左手拇指在滴定管靠近身体一侧，食指在滴定管远离身体一侧，捏住乳胶管中的玻璃球所在部位稍上处并向手心挤捏乳胶管，使其与玻璃球之间形成一条缝隙，溶液即可流出。应注意，不能捏挤玻璃球下方的乳胶管，否则会进入空气形成气泡。为防止乳胶管来回摆动。可用中指和无名指夹住尖嘴的上部（图2-9）。

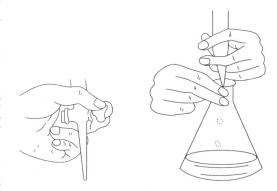

图2-9 滴定管操作示意图

在使用酸碱通用滴定管时，其操作方法与酸式滴定管操作方法类似。

滴定通常在锥形瓶内进行，必要时也可以在烧杯中进行。对于滴定碘法、溴酸钾法等，则需在碘量瓶中进行反应和滴定。滴定结束后，把滴定管中剩余的溶液倒掉（不能倒回原贮液瓶），依次用自来水和蒸馏水清洗，然后用蒸馏水充满滴定管并垂直夹在滴定管架上。酸性滴定管长时间不用时，需用纸片夹在磨口塞之间，防止磨口塞粘连（图2-9）。

（三）容量瓶

容量瓶是一种具塞细颈梨形平底瓶。瓶颈上刻有环形标线，表示在瓶身标识温度下（一般为20℃）液体充满至标线时的容积。容量瓶用来把精密称量的物质配制成准确浓度的溶液，也可将已知准确浓度的浓溶液稀释成稀溶液。通常容量瓶有10、25、50、100、250、500及1000ml等规格。

1. 检漏 首先检查瓶塞与瓶身磨口是否密封。将容量瓶内注入适量自来水，盖好瓶塞，右手托住瓶底，将其倒立2分钟，观察瓶塞周围是否有水渗出。如果不漏，再把塞子旋转180°，塞紧并倒置，如仍不漏水，则可使用。容量瓶与瓶塞要配套使用，若瓶塞与容量瓶磨口不匹配，会造成漏液。因此瓶塞必须用细绳系在瓶颈上，以防丢失或摔碎。

2. 清洗 容量瓶使用前需用铬酸洗液浸泡，之后依次用自来水、蒸馏水冲洗，清洗后内壁不挂水珠视为合格。

3. 配制溶液 精确称量试剂置于烧杯中，加入适量蒸馏水溶解。用玻璃棒辅助转移溶液至容量瓶中。烧杯中的溶液倒尽后烧杯不要直接离开玻璃棒，而应在烧杯扶正的同时使杯嘴沿玻璃棒上提1~2cm，随后烧杯即可离开玻璃棒，这样可避免杯嘴与玻璃棒之间的一滴溶液流到烧杯外面。然后用少量蒸馏水洗涤烧杯壁3~4次，每次的洗液按照同样操作转移至容量瓶中。当溶液达到容量瓶的2/3容量时，应将容量瓶沿水平方向摇晃，使溶液初步混匀（注意，不能倒转容量瓶！）再加水至距离标线2~3cm处，改用滴管缓慢滴加蒸馏水直至溶液弯月面最低点恰好与标线相切。盖紧瓶塞，并用食指将其压住，另一只手五指托住容量瓶底，倒转容量瓶，使瓶内气泡上升到顶部，边倒转边向同一方向摇动。如此反复倒转摇动多次，直至瓶内溶液充分混匀（图2-10）。

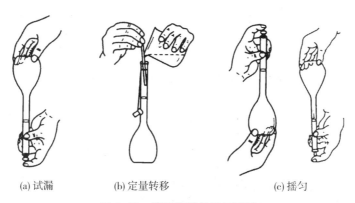

(a) 试漏　　(b) 定量转移　　(c) 摇匀

图 2-10　容量瓶的使用示意图

容量瓶不宜配制更不宜贮藏强碱性溶液，如溶液需要使用较长时间，应将溶液转移至试剂瓶中贮存。试剂瓶应先用该溶液润洗2~3次，以保证浓度不变。容量瓶用后应及时清洗，不得将容量瓶置于烘箱中烘烤，也不能以任何方式对其加热。容量瓶长时间不用时应在瓶塞和瓶身的磨口间插入纸条，防止磨口粘连。

（四）移液管、吸量管

移液管和吸量管是结构相似、用途相同的玻璃量器，都是用于准确移取一定体积的溶液。移液管是中间有膨大部分（称为球部）的长玻璃管，球部上下均为较细的玻璃管颈，在上端玻璃管上刻有一条标线。常用的规格有1、2、5、10、25、50ml等。吸量管与移液管相比，没有球部，在其管身上刻有分刻度，用它可以准确移取溶液。一般吸量管的规格比移液管稍小，常用的规格有1、2、5、10ml等。

移液管在使用前需用铬酸洗液浸泡数小时，再用自来水、蒸馏水冲洗，最后用滤纸吸干移液管尖嘴处的水。使用前用待移取的溶液润洗2~3次，以确保所移取溶液的浓度不变。

移取溶液时，右手大拇指和中指捏住管身上端，尖嘴插入溶液中。左手持洗耳球，先把球中空气挤出，然后将洗耳球尖端插入移液管上端管口，缓慢松开洗耳球，利用大气压力将溶液吸入移液管。当液面达到目标刻度线以上后，移去洗耳球，立即用右手的食指按住移液管管口，将移液管尖嘴提出液面，靠在盛放溶液器皿的内壁上，略微放松右手食指，用右手大拇指和中指轻轻捻转管身，使液面平稳下降。当溶液的弯月面与标线相切时，食指压紧管口，使液体不再流出（图2-11）。

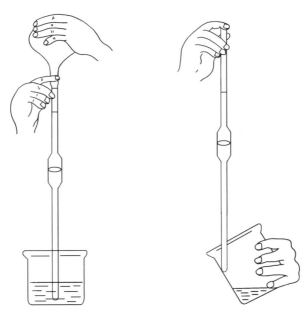

图 2-11 移液管（或吸量管）的操作示意图

取出移液管，插入盛接溶液的器皿，使管的尖嘴靠在器皿内壁上。移液管保持竖直，盛接器皿略倾斜，松开右手示指，让管内溶液自然地沿器壁流下。溶液完全流出后，移液管尖嘴继续靠壁捻转管身 10~15 秒再拿出。

若移液管未标"吹"字，则不可用外力使残留在移液管末端的溶液流出。有一些容量较小的吸量管，如 0.1ml，管口上刻有"吹"字，在使用时，末端的溶液需要吹出，否则将造成误差。实验时，应尽量使用同一支吸量管进行量取，以避免误差。

六、溶液的配制

在此主要介绍一般溶液和标准溶液的配制方法。

（一）一般溶液的配制方法

一般溶液也称辅助试剂试液。配制该溶液时，对浓度的准确度要求不高，一般只需保留 1~2 位有效数字。一般溶液的配制方法主要包括三种：直接水溶法、介质水溶法、稀释法。

1. 直接水溶法　该方法主要适用于易溶于水且不发生水解或者水解程度较小的固体试剂，例如：氢氧化钠、氯化钠等。利用此方法配制溶液时，首先利用电子天平称量一定量的试剂，并转移至小烧杯中，然后加入少量蒸馏水使试剂全部溶解，再用蒸馏水稀释至所需体积，最后混匀后转移至试剂瓶中，并贴好标签。

2. 介质水溶法　该方法主要适用于易发生水解反应的固体试剂，例如：三氯化铁、氯化亚锡等。为避免这类试剂发生水解反应，因此在配制溶液时，需先加入一定量的介质，即一定浓度的酸溶液或者碱溶液。具体配制过程如下：首先利用电子天平称取一定量的试剂，并转移至小烧杯中，然后加入适量一定浓度的相应的酸溶液或者碱溶液，使试剂全部溶解，再用蒸馏水稀释至所需体积，最后混匀后转移至试剂瓶中，并贴好标签。

3. 稀释法　该方法主要适用于浓溶液的稀释，例如：浓硫酸、浓盐酸、浓硝酸等。一般配制过程如下：首先利用量筒量取一定体积的浓溶液，并转移至小烧杯中，然后用蒸馏水稀释至所需体积，最后混匀后转移至试剂瓶中，并贴好标签。但需强调的是：利用浓硫酸配制稀硫酸时，应在不断搅拌下将浓硫酸缓慢加入到蒸馏水中，切不可将加入顺序颠倒！

（二）标准溶液的配制方法

标准溶液的配制方法主要有两种：直接配制法和间接配制法。

1. 直接配制法 称取一定量的基准物质，溶解后定量转移到容量瓶中，稀释至刻度，根据所称取基准物质的质量和容量瓶的体积，即可计算出标准溶液的准确浓度，这种方法仅适用于基准物质，例如重铬酸钾、碳酸钠等。

具体配制过程如下：首先将准确称量的药品放入干净的小烧杯中，加入少量溶剂，利用玻璃棒搅拌至药品完全溶解；然后利用玻璃棒将烧杯中溶液引流转移至容量瓶中，再用溶剂将小烧杯润洗3~4次，并将每次的润洗液用同样的方法转移至容量瓶，保证所有药品转移至容量瓶中；接着向容量瓶中加入溶剂，直至达到容量瓶体积的2/3左右，再将容量瓶平行于水平面摇动几圈（切勿倒转容量瓶），使溶液初步混匀；随后继续向容量瓶中加入溶剂，当液面接近标线后，将容量瓶静置1分钟，此后改用滴管将溶剂滴加到容量瓶中，直至溶液弯液面下缘恰好与标线相切；最后，将瓶塞盖紧，左手食指按压瓶塞，右手五指指腹托住容量瓶底部，倒转并旋摇容量瓶，使瓶内气泡上升至顶部，如此反复几次，使容量瓶内溶液混合均匀。

2. 间接配制法 许多物质由于达不到基准物质的要求，只能采用间接法配制，即先粗略称取一定量物质配制成接近于所需浓度的溶液，再用基准物质或另一种已知准确浓度的标准溶液来测定其准确浓度。这种利用基准物质或者已知准确浓度的标准溶液来测定待测标液浓度的操作过程称为标定。例如，配制氢氧化钠溶液等。

七、滤纸及试纸的使用

（一）滤纸的使用

在化学实验中，滤纸常被用作过滤介质，使溶液与固体分离。目前，常用的滤纸主要分为定量滤纸和定性滤纸。每种滤纸按孔隙大小又可分为快速、中速、慢速三种。除了做沉淀的质量分析外，其余过滤实验一般选用定性滤纸作为过滤介质。在选用滤纸时，应根据沉淀类型进行确定。其中，细晶形沉淀主要选用慢速滤纸；粗晶形沉淀可选用中速滤纸；胶状沉淀可选用快速滤纸。

利用滤纸进行常压过滤时具体的使用方法如下（图2-12）：取一尺寸合适的圆形滤纸，对折两次后展开得到圆锥形，并且使圆锥形的一边是三层，另一边为一层，此时圆锥形内角为60°，恰好能与漏斗内壁贴合；为了使滤纸上沿与漏斗壁紧贴而无气泡，常将三层滤纸的最外面两层撕去一个小角；然后用食指将滤纸按在漏斗内壁中，用少量去离子水润湿滤纸，并用玻璃棒轻压滤纸四周，去除滤纸与漏斗壁间的气泡，使滤纸紧贴在漏斗壁上；最后在漏斗中加水至滤纸边缘，此时漏斗颈内应全部充满水，形成水柱。

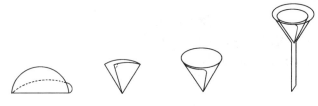

图 2-12 滤纸的折叠与放置

滤纸使用注意事项如下。

（1）若漏斗的内径角度大于或者小于60°，应适当增大或者减小滤纸折成的角度，使滤纸与漏斗壁贴合。

（2）滤纸边缘要比漏斗口略低0.5~1cm。

（3）若在漏斗颈中未形成水柱，可用手指堵住漏斗下口，稍掀起滤纸的一边用洗瓶向滤纸和漏斗间加水，直到漏斗颈和锥体部分被水充满，而且漏斗颈内气泡完全排出，最后，轻压滤纸边，松开堵住漏斗的手指，即可形成水柱。

（二）试纸的使用

在化学实验中，试纸常被用于快速、定性检验一些溶液或气体的酸碱性或某些物质是否存在。目前，常见的试纸有石蕊试纸、pH试纸、醋酸铅试纸、淀粉–碘化钾试纸等。各种试纸的适用范围和使用方法如下。

1. 石蕊试纸 主要用于定性检验溶液或者气体的酸碱性。根据检测性质，该试纸主要分为两类：蓝色石蕊试纸和红色石蕊试纸。其中，蓝色石蕊试纸用于检测pH<5的溶液或者酸性气体，蓝色石蕊试纸遇上述酸性物质变成红色；红色石蕊试纸用于检测pH>8的溶液或者碱性气体，红色石蕊试纸遇上述碱性物质变成蓝色。

利用石蕊试纸检测溶液酸碱性时，首先将试纸剪成小块，然后用镊子将一小块试纸放置于干净且干燥的表面皿或者白色点滴板上，最后将玻璃棒蘸取的待测溶液点在试纸中部，观察试纸颜色变化。若待检测物为气体时，则需先将试纸用蒸馏水润湿，然后再用镊子将试纸放置于气体出口处，观察试纸颜色变化。

2. pH试纸 主要用于测量溶液pH，该试纸遇到酸碱性强弱不同的溶液时会显示不同的颜色。根据可测量的pH范围，该试纸主要分为两类：广泛pH试纸和精密pH试纸。其中，广泛pH试纸的变色范围为pH 1~14，可粗略测量溶液pH。精密pH试纸的变色范围则相对较小，可较精确的测量溶液的pH。常见的精密pH试纸变色范围包括2.7~4.7、3.8~5.4、5.4~7.0、6.9~8.4、8.2~10.0、9.5~13.2。

pH试纸的使用方法与石蕊试纸的使用方法基本相同。唯一不同的是，在试纸发生颜色变化后需要与标准色板进行比较，从而确定溶液的pH。

3. 醋酸铅试纸 主要用于定性检验H_2S气体。其检验原理是基于醋酸铅可与H_2S发生化学反应生成黑色的PbS，从而使试纸变成黑褐色且有金属光泽。

利用醋酸铅试纸检验H_2S气体时，首先将试纸用去离子水润湿，然后用镊子将试纸放置于气体出口处，观察试纸是否变成黑褐色且有金属光泽。

4. 淀粉–碘化钾试纸

淀粉–碘化钾试纸主要用于定性检验氧化性气体，例如Cl_2、Br_2。其检验原理是基于氧化性气体可以将试纸中的I^-氧化为I_2，I_2与试纸中淀粉作用后生成蓝紫色物质，从而使试纸变成蓝紫色。

利用淀粉–碘化钾试纸检验氧化性气体时，首先将试纸用去离子水润湿，然后用镊子将试纸放置于气体出口处，观察试纸是否变成蓝紫色。

5. 试纸使用注意事项

（1）试纸需要密闭保存，取试纸需用镊子，防止污染。

（2）测定溶液时，不能将试纸直接插入待测溶液中。

（3）测定气体时，不能将试纸与容器口接触。

八、沉淀的分离和洗涤

（一）沉淀与溶液的分离

溶液与沉淀（或结晶）的分离方法一般有倾析法、离心分离法和过滤法三种。

1. 倾析法 当沉淀的相对密度较大或结晶颗粒较大且静置后能较快地沉降至容器的底部时，可用倾析法进行沉淀的分离和洗涤。把沉淀上部的澄清溶液倾入另一容器内，然后加入少量洗涤液（如蒸馏水）洗涤沉淀，充分搅拌沉降，倾去洗涤液。如此重复操作 2~3 次，即可将沉淀洗净。

2. 离心分离法 少量溶液和沉淀物分离时，采用离心分离法。此法简便、快速。如在试管反应中，用一般的过滤法，沉淀黏在滤纸上难以取下，不便于进一步的实验。一般来说，当沉淀的相对密度小于"1"时，不能用离心分离法分离。实验室中常用的离心仪器是电动离心机（图 2-13）。

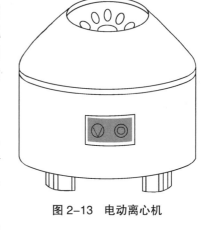

图 2-13　电动离心机

离心机使用注意事项：

（1）离心管放入金属导管中，位置要对称，重量要平衡，否则易损坏离心机的轴。如果只有一支离心管的沉淀需要进行分离，可取另一支规格相同的、空的离心管，盛以相应质量的水，然后把离心管分别对称地装入离心机的套管中，以保持平衡。

（2）打开旋钮，逐渐旋转变阻器，使离心机转速由小到大。数分钟后慢慢恢复变阻器到原来的位置，使其自行停止。

（3）离心时间和转速由沉淀的性质来决定。结晶形的紧密沉淀，转速每分钟 1000 转，1~2 分钟后即可停止。无定形的疏松沉淀，沉降时间要长些，转速可提高到每分钟 2000 转。如果经 3~4 分钟后仍不能使其分离，则应设法（如加入电解质或加热等）促使沉淀沉降，然后再进行离心分离。

离心分离操作步骤如下。

（1）沉淀　边搅拌溶液边加沉淀剂，等反应完全后，离心沉降。在上层清液中再加试剂一滴，如清液不变浑浊，即表示沉淀完全，否则必须再加沉淀剂直至沉淀完全，再离心分离。

（2）溶液的转移　离心沉降后，用吸管把清液与沉淀分开。其方法是，先用手指捏紧吸管上的橡皮头，排除空气，然后将吸管轻轻插入清液（切勿在插入清液以后再捏橡皮头），慢慢放松橡皮头，溶液即慢慢进入吸管中，随试管中溶液的减少，将吸管逐渐下移至全部溶液吸入管内为止。吸管尖端接近沉淀时要特别小心，勿使其触及沉淀（图 2-14）。

图 2-14　溶液与沉淀分离

（3）沉淀的洗涤　如果要将离心后的沉淀溶解后再做鉴定，必须在溶解之前，将沉淀洗涤干净。常用的洗涤剂是蒸馏水。加洗涤剂后，用搅拌棒充分搅拌，离心分离，清液用吸管吸出，必要时可重复洗几次。

3. 过滤 过滤法是将溶液与沉淀分离最常用的方法。过滤时，溶液与沉淀的混合物通过过滤器如滤纸），沉淀留在过滤器上，溶液则通过过滤器进入盛接的容器中，所得溶液称为滤液。溶液的温度、黏度、过滤时的压力，过滤器孔隙的大小和沉淀物的性质都会影响过滤的速度。化学实验中常用的过滤方法有常压过滤法、减压过滤法、热过滤法和倾析法过滤法四种。

（1）常压过滤　在常压下用普通漏斗过滤的方法称为常压过滤法，此法最为简便和常用。过滤器是玻璃漏斗和滤纸。当沉淀物为胶体或细微晶体时，用此法过滤较好，缺点是过滤速度较慢。这是一种最简单和常用的过滤方法，操作步骤如下。

滤纸的折叠：取一正方形或圆形滤纸折叠成四层并剪成扇形，圆形滤纸不必再剪。若漏斗的规格不标准（非 60° 角），滤纸和漏斗不密合，这时需要重新折叠滤纸，不对半折而成一个适当的角度，展开后可以展成大于 60° 的锥形，也可展成小于 60° 的锥形，根据漏斗的角度来选用，使滤纸与漏斗密合，然后撕去一小角。

用食指把滤纸按在漏斗内壁上，用水湿润滤纸，并使它紧贴在壁上，去除滤纸和漏斗内壁之间的气泡。过滤时，漏斗颈内可充满滤液，滤液以本身的重量使漏斗内液下漏，过滤大为加速，否则，气泡的存在可阻碍液体在漏斗颈内流动而减缓过滤的速度。漏斗中滤纸的边缘应略低于漏斗的边缘（图2-15）。

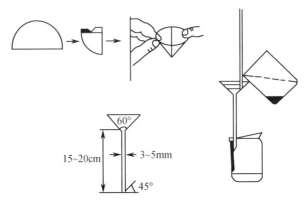

图2-15　滤纸的折叠方法和过滤操作

过滤注意事项如下。

① 漏斗放在漏斗架上，并调整漏斗架的高度，使漏斗的出口靠在接受容器的内壁上，以便使溶液顺着容器壁流下，减少空气阻力，加速滤程，且防止滤液溅出。

② 将溶液转移到漏斗中时，要采用倾析法。先倾倒溶液，后转移沉淀，这样就不会因为沉淀堵塞滤纸的孔隙而减慢过滤速度。

③ 转移溶液时，应使用玻璃棒引流，让溶液顺其缓缓倾入漏斗中，玻璃棒下端轻轻触在三层滤纸处。注意不能触在单层滤纸处，以免把单层滤纸捅破。

④ 过滤过程中，漏斗中的溶液不能太多，液面应低于滤纸上缘3~5mm，以防过多的溶液沿滤纸和漏斗内壁的缝隙流入接收器，失去滤纸的过滤作用。

（2）减压过滤　减压过滤又称抽滤或真空过滤。减压可以加快过滤的速度，还可以把沉淀抽吸得比较干燥。但不宜用于过滤胶状沉淀和颗粒太小的沉淀。因为胶状沉淀在快速过滤时易透过滤纸，而颗粒太小的沉淀易在滤纸上形成一层密实的沉淀致使溶液不易透过，装置如图2-16所示。

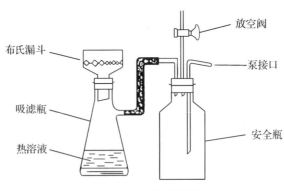

图2-16　减压过滤装置

操作时应该注意以下几点：

① 过滤前须检查漏斗的颈口是否对准吸滤瓶的支管，安全瓶的长玻璃管接水泵，短的接吸滤瓶。

② 滤纸的大小应剪得恰好掩盖住漏斗的磁孔，先用水或相应的溶剂润湿，然后开启水泵，使

它贴紧漏斗不留孔隙,这时才能进行过滤操作。

③ 过滤时,先将上部澄清液沿着玻璃棒注入漏斗中,然后再将晶体或沉淀转入漏斗进行吸滤。未能完全转移的固体应用母液冲洗再行转移而不能用水或相应的溶剂,以减少沉淀的损失。

④ 滤液将充满吸滤瓶时(但不能使它上升至吸滤瓶支管的水平位置),应拔去橡皮管,停止抽气,将漏斗拿下,将滤液从吸滤瓶中倒出后再继续抽滤。

⑤ 在抽滤过程中,不得突然关闭水泵,如欲取出沉淀或是倒出滤液而需要停止抽滤时,应该先将吸滤瓶支管上橡皮管拔下,停止抽滤,然后再关上水泵,否则水将倒吸。

⑥ 在漏斗内洗涤结晶时,应停止抽滤,让少量水或相应的溶液缓慢通过晶体,然后再行抽滤和压干。

有些强酸性、强碱性或强氧化性的溶液过滤时不能用滤纸,因为溶液要和滤纸作用而破坏滤纸,可用石棉纤维来代替滤纸。此法适用于分析或滤液有用的情况。还有使用玻璃熔砂漏斗的,这种漏斗常见的规格有四种,即1号、2号、3号、4号。1号的孔径最大。可以根据沉淀颗粒不同来选用。但它不能用于强碱性溶液的过滤,因为强碱会腐蚀玻璃。

(3)热过滤　如果溶液中的溶质在温度下降时很易析出大量结晶,为不使结晶在过滤过程中留在滤纸上,就要趁热进行过滤。过滤时可把玻璃漏斗放在铜质的热漏斗内(图2-17),热漏斗内装有热水,以维持溶液的温度。

也可以在过滤前把玻璃漏斗放在水浴上用蒸汽加热,然后使用,此法较简单易行。另外,热过滤时选用的玻璃漏斗的颈部愈短愈好,以免过滤时溶液在漏斗颈内停留过久,因散热降温,析出晶体而发生堵塞。

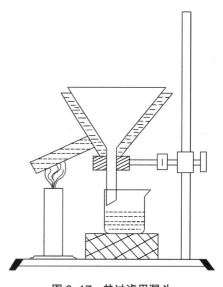

图2-17　热过滤用漏斗

(4)倾析过滤法　过滤前,先让沉淀沉降;过滤时,不要搅动沉淀,先把清液倒入滤纸上,待清液滤完,再把沉淀转移到滤纸上。这样可防止沉淀堵塞滤孔而减慢过滤速度。最后由洗瓶吹出少量蒸馏水,洗涤沉淀1~2次。需要充分洗涤沉淀时,还可在倾出清液后,用蒸馏水洗涤,重复数次。

(二)沉淀的洗涤

洗涤沉淀是为了洗去沉淀表面吸附的杂质和混杂在沉淀中的母液。洗涤时要尽量减少沉淀的溶解损失和避免形成胶体,因此需选择合适的洗液。选择洗液的原则:对于溶解度很小且又不易形成胶体的沉淀,可用蒸馏水洗涤;对于溶解度较大的晶体沉淀,可用沉淀剂稀溶液洗涤,但沉淀剂必须在烘干或灼烧时易挥发或易分解而被除去;对于溶解度较小而又可能分散成胶体的沉淀,应用易挥发的电解质稀溶液洗涤。

用热洗涤液洗涤,则过滤较快,且能防止形成胶体,但溶解度随温度升高而增大较快的沉淀不能用此方法。

洗涤必须连续进行,一次完成,不能将沉淀干涸放置太久,尤其是一些非晶形沉淀,放置凝聚后,不易洗净。

洗涤沉淀时,既要将沉淀洗净,又不能增加沉淀的溶解损失。用适当少的洗液,分多次洗涤,可以提高洗涤效果。

九、溶解、蒸发和结晶

（一）溶解

1. 固体的研磨　若固体物质颗粒较大，溶解之前，往往需要进行粉碎。实验室中粉碎固体一般在研钵中进行。用研杵在研钵中将固体物质磨成细小颗粒或粉末，能加速固体溶解。

研磨操作注意事项：

（1）研磨物质的体积不得超过钵体容积的1/3。

（2）研磨时用研杵将固体颗粒挤压到研钵内壁，进行转圈研磨，不能用研杵敲击固体。

（3）易燃、易爆和易分解的物质不能用研磨的方法粉碎。

实验室中的研磨设备除了研钵，还有较高级的小型球磨机和胶体磨等，可以将固体物质颗粒直径磨细到1~5μm。

2. 溶解　用溶剂溶解固体试样时，加入溶剂时应先把烧杯适当倾斜，然后把量筒嘴靠近烧杯壁，让溶剂慢慢顺着杯壁流入；或通过玻璃棒使溶剂沿玻璃棒慢慢流入，以防杯内溶液溅出而损失。溶剂加入后，用玻璃棒搅拌，使试样完全溶解。对溶解时会产生气体的试样，则应先用少量水将其润湿成糊状，用表面皿将烧杯盖好，然后用滴管将溶剂自杯嘴逐滴加入，以防生成的气体将粉状的试样带出。对于需要加热溶解的试样，加热时要盖上表面皿，要防止加热时溶液剧烈暴沸而溅出，加热和搅拌可加快溶解速度。加热后要用蒸馏水冲洗表面皿和烧杯内壁，冲洗时也应使水顺杯壁流下。

在实验的整个过程中，盛放试样的烧杯要用表面皿盖上，以防杂质落入。放在烧杯中的玻璃棒，不要随意取出，以免溶液损失。

（二）蒸发

为鉴定含有较少的离子，在鉴定前应将溶液浓缩。溶液的浓缩一般在小烧杯中进行。烧杯放在石棉网中央，手持煤气灯以小火在下面来回移动使溶液蒸发缓慢均匀，不致因溅出而损失。

若须蒸发至干时，应在蒸发近干即停止加热，让残液依靠余热自行蒸干，避免固体溅出，同时也可防止物质分解。

有时，溶液蒸干后所留下的固体若需强热灼烧，在这种情况下，溶液的蒸发应在小坩埚中进行，蒸发方法与前相同。蒸干后放在小的泥三角上用火烘干，加热的火焰开始小些，然后逐渐加大火焰直至炽热灼烧。

溶液的蒸发浓缩通常在蒸发皿中进行。在少数情况下亦可在烧杯中加热蒸发浓缩，但蒸发效率较差。应用蒸发皿蒸发浓缩溶液时应注意下列几点。

（1）蒸发器内所放液体的体积不应超过容量的2/3。

（2）蒸发溶液应缓慢进行，不能加热至沸腾。

（3）蒸发溶液应在水浴锅上进行（少数情况下可放在石棉网上加热），不可用火直接加热。

（4）蒸发过程中应不断用搅拌棒刮下由于体积缩小而留于液面边缘上的固体。

（5）溶液浓缩程度随溶质溶解度大小而不同，但应尽量避免溶液蒸至干涸。

（6）由蒸发皿倒出液体应从嘴沿搅拌棒倒出。

（三）结晶

各种晶体都有特征的晶形。影响晶体生长的因素很多，这些因素不仅会影响结晶速度及晶体大小，有时还会改变结晶的形状。所以要得到一定形状的晶体，要有合适的结晶条件。一般来讲，由较稀的溶液中得到的晶体较大，晶形较好；而由较浓的溶液中得到的晶体较细，晶形不易完整。

1. 显微结晶反应　由于各种晶体都有特征的晶形，故可用显微镜观察反应生成的晶体形状，并很快地做出某种离子是否存在的结论。

显微结晶反应的操作方法如下：在干燥的显微镜载片上，相距2cm左右各滴试液与试剂一滴，然后用细的玻璃棒沟通，使试剂与试液发生缓慢的反应，结果在中间先生成晶体。观察晶形时，应将过多的溶液用滤纸吸去。

如果溶液浓缩后才能结晶，则必须使溶液在载片上受热蒸发。操作方法是：先滴一滴试液于载片的中央，然后用试管夹夹载片的一端在石棉网的上方来回移动使其受热，缓慢蒸发至干，冷却后在残渣上加一滴试剂，过一些时间就会生成晶体。

观察生成的晶体须用显微镜。

使用显微镜的方法如下。

（1）选择放大倍数合适的目镜及物镜（放大总倍数为目镜和物镜放大倍数之乘积）。

（2）调好反光镜，使目镜内照明良好。

（3）在载物台上放好载片。载片背面应擦干，以免污染物台。载片应夹好以防滑动。

（4）调节物镜最低至离载片5mm左右，然后用左眼看目镜并缓慢升高镜筒，直至呈现清晰的物像为止。若镜筒升至最高仍未看到现象，应重新将镜筒降至离载片5mm后再重新调节，绝不可在观察时下降镜筒，以防物镜触及载片。

（5）目镜及物镜若被污染，应当用擦镜纸，不能用一般的纸或布擦。

（6）显微镜不用时应放在箱内，物镜放在专用盒中。

2. 重结晶　从混合物中分离出的固体化合物往往是不纯的，其中常夹杂一些反应副产物、未作用的原料及催化剂等。纯化这类化合物的有效方法通常是用合适的溶剂进行重结晶。其一般过程如下。

（1）将不纯的固体有机物在溶剂的沸点或接近沸点的温度下溶解在溶剂中，制成接近饱和的浓溶液，若固体有机物的熔点较溶剂沸点低，则应制成在熔点温度以下的饱和溶液。

（2）若溶液含有色杂质，可加活性炭煮沸脱色。

（3）过滤此热溶液以除去其中不溶物质及活性炭。

（4）将滤液冷却，使结晶自过饱和溶液中析出，而杂质仍留母液中。

（5）提取和过滤。从母液中将结晶分出，洗涤结晶以除去吸附的母液。所得的结晶，经干燥后测定熔点，如发现其纯度不符合要求时，可重复上述操作直至熔点不再改变。

3. 溶液结晶　将滤液在冷水浴中迅速冷却并剧烈搅动时，可得到颗粒很小的晶体。小晶体中包含杂质较少，但其表面积较大，吸附于其表面的杂质较多。若希望得到均匀而较大的晶体，可将滤液（如在滤液中已析出结晶，可加热使之溶解）在室温或保温下静置使之缓缓冷却。

有时由于滤液中焦油状物质或胶状物存在，使结晶不易析出，或有时因形成过饱和溶液也不析出结晶，在这种情况下，可用玻璃棒摩擦器壁以形成粗糙面，使溶质分子呈定向排列而形成结晶的过程较在平滑面上迅速和容易；或者投入晶种（同一物质的晶体，若无此物质的晶体，可用玻璃棒蘸一些溶液稍干后即会析出结晶），供给定型晶核，使晶体迅速形成。

有时被纯化的物质呈油状析出，油状物长时间静置或足够冷却后虽也可以固化，但这样的固体往往含有较多杂质（杂质在油状物中溶解度通常比在溶剂中溶解度大；其次，析出的固体中还会包含一部分母液），纯度不高，用溶剂大量稀释，虽可防止油状物的生成，但将使产物大量损失。这时可将析出油状物的溶液加热重新溶解，然后慢慢冷却。一旦油状物析出时便剧烈搅拌混合物，使油状物在均匀分散的状况下固化，这时包含的母液就大大减少。但最好还是重新选择溶剂，得到有晶形的产物。

十、酸度计的使用

酸度计（也称 pH 计）是用来测量溶液 pH 的仪器，还可用于测量电池电动势 (mV)。酸度计主要由参比电极（饱和甘汞电极，图 2-18）、测量电极（玻璃电极，图 2-19）和精密电位计组成；复合电极则是参比电极和测量电极合在一起制成的复合体。实验室常用的酸度计有 pHS-2 型和 pHS-3C 型等多种型号，各种型号结构略有差别，但基本原理相同。

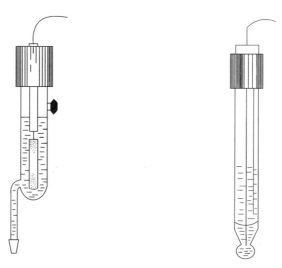

图 2-18　饱和甘汞电极示意图　　图 2-19　玻璃电极示意图

（一）基本原理

直接电位法测定溶液 pH 常用玻璃电极作为指示电极（负极），饱和甘汞电极作为参比电极（正极），浸入待测溶液中组成原电池：

（−）Ag|AgCl(s), 内充液|玻璃膜|试液 ‖ KCl(饱和), $Hg_2Cl_2(s)$|Hg（＋）

此原电池电动势为：$E_{MF} = E_甘 - E_玻 = K' + \dfrac{2.303RT}{F}\text{pH}$

由上式可见，原电池的电动势与溶液 pH 呈线性关系，斜率为 $2.303RT/F$，它是指溶液 pH 变化一个单位时，电池的电动势变化 $2.303RT/F$（V）（25℃ 时改变 0.059V）。为了直接读出溶液的 pH，pH 计上相邻两个读数间隔相当于 $2.303RT/F$ (V) 的电位，此值随温度的改变而变化，因此 pH 计上均设有温度调节旋钮，以消除温度对测定的影响。

上式中 K' 受诸多不确定因素影响，难以准确测定或计算得到，因此在实际测量时，常采用"两次测量法"进行，首先用已知 pH_S 的标准缓冲溶液来校准 pH 计，称为"定位"，使电池电动势 E_S 和溶液 pH_S 的关系能满足上式，然后再在相同条件下测量待测液的 pH_X，这样可消除 K' 的影响，因此待测溶液 pH_X 表示为：

$$pH_X = pH_S + \dfrac{E_X - E_S}{0.059}\ (25℃)$$

由此可见，pH 测量是相对的，每次测量均需与标准缓冲溶液进行对比，因此测量结果的准确度受标准缓冲溶液 pH_S 值准确度的影响。

（二）pHS-3C 型 pH 计的使用

pHS-3C 型 pH 计的结构见图 2-20。

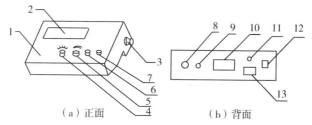

1-面板，2-显示屏，3-电极梗插座，4-温度补偿调节旋钮，5-斜率补偿调节旋钮，
6-定位调节旋钮，7-选择旋钮（pH 或 mV），8-测量电极插座，9-参比电极插座，
10-铭牌，11-保险丝，12-电源开关，13-电源插座

图 2-20 pHS-3C 型 pH 计示意图

1. 开机，安装电极 将电源线插入电源插座 13，按下电源开关 12，接通电源后需预热 30 分钟。将电极梗插入电极梗插座 3，电极夹夹在电极梗上，取下复合电极前端的电极套，将电极夹在电极夹上。

2. 标定 仪器使用前，要先标定。一般情况下，仪器连续使用时每天要标定一次。标定方法如下。

（1）拔下测量电极插座 8 处的短路插头，插上复合电极。

（2）把选择旋钮 7 调到 pH 档。

（3）调节温度旋钮 4 至待测溶液温度值。

（4）调节斜率旋钮 5 至 100% 位置。

（5）用蒸馏水清洗电极，并用滤纸吸干后插入 pH=6.86 的标准缓冲溶液中。按溶液温度查出该温度时缓冲溶液 pH，调节定位旋钮，使仪器显示读数与该缓冲溶液的 pH 一致。

（6）清洗电极，并用滤纸吸干。若被测溶液为酸性，则再用 pH=4.00 的标准缓冲溶液调节斜率旋钮使仪器显示读数为 4.00；若被测溶液为碱性，则再用 pH=9.18 的标准缓冲溶液调节斜率旋钮使仪器显示读数为 9.18。

经标定的仪器，测量时不得再转动定位调节旋钮和斜率调节旋钮。

3. pH 测量 将纯化水洗净的电极插入被测溶液中，轻轻摇动烧杯，使溶液均匀，静止后，直接在显示屏上读出溶液的 pH。

一般若测定偏碱性溶液时，应用 pH 6.86 和 pH 9.18 标准缓冲溶液来校正仪器；测定偏酸性溶液时，则用 pH 4.00 和 pH 6.86 的标准缓冲溶液。校正时标准溶液的温度与状态 (静止还是流动) 应尽量和被测液的温度与状态一致。在使用过程中，如遇到更换新电极或"定位"或"斜率"调节器变动过的情况时，仪器必须重新校正。

4. 电池电动势（mV）的测量 将指示电极、参比电极和电源分别插入相应的插座中，把选择调节器调到 mV 位置，将电极系统放入被测溶液中，即可进行测定。

5. 四种标准缓冲溶液的配制

（1）草酸三氢钾标准缓冲溶液（0.05mol/L） 精密称取在（54±3）℃ 干燥 4~5 小时的草酸三氢钾 [$KH_3(C_2O_4)_2 \cdot 2H_2O$] 12.71g，加水使其溶解并稀释至 1000ml。

（2）邻苯二甲酸氢钾标准缓冲溶液（0.05mol/L） 精密称取在 (115±5)℃ 干燥 2~3 小时的邻苯二甲酸氢钾（$KHC_8H_8O_4$）10.21g，加水使其溶解并稀释至 1000ml。

（3）磷酸盐标准缓冲溶液（0.025mol/L） 精密称取在（115±5）℃ 干燥 2~3 小时的磷酸氢二钠（Na_2HPO_4）3.55g 和磷酸二氢钾（KH_2PO_4）3.40g，加水使其溶解并稀释至 1000ml。

（4）硼砂标准缓冲溶液（0.01mol/L） 精密称取硼砂（$Na_2B_4O_7 \cdot 10H_2O$）3.81g（注意避免风化），加水使其溶解并稀释至 1000ml，置聚乙烯塑料瓶中，密闭保存。

注：配制标准缓冲溶液的水，应是新煮沸过并放冷的纯水；标准缓冲溶液一般可保存 2~3 个

月，但发现有浑浊、沉淀或霉变等现象时，不能继续使用。

6. 标准缓冲溶液的pH与温度（0~50℃）关系对照表

温度 （℃）	草酸三氢钾 （0.05mol/L）	邻苯二甲酸氢钾 （0.05mol/L）	KH$_2$PO$_4$+Na$_2$HPO$_4$ （0.025mol/L）	硼砂 （0.01mol/L）
0	1.666	4.003	6.984	9.464
5	1.668	3.999	6.951	9.395
10	1.670	3.998	6.923	9.332
15	1.672	3.999	6.900	9.276
20	1.675	4.002	6.881	9.225
25	1.679	4.008	6.865	9.180
30	1.683	4.015	6.853	9.139
35	1.688	4.024	6.844	9.102
38	1.691	4.030	6.840	9.081
40	1.694	4.035	6.838	9.068
45	1.700	4.047	6.834	9.038
50	1.707	4.060	6.833	9.011

7. 复合电极的使用和维护

（1）取下电极保护套后，要避免电极的敏感玻璃泡与硬物接触，以免因破损或磨损而使电极失效。

（2）测量前必须用已知pH的标准缓冲溶液进行标定，要保证缓冲溶液的可靠性。

（3）测量后及时将电极套套上，套内放入少量补充液以保持电极玻璃球泡湿润。

（4）电极避免长期浸泡在蒸馏水、蛋白质溶液或酸性氟化物溶液中，避免与有机硅油接触。

（5）电极长期使用后，若斜率略有降低，可将电极下端浸入4% HF中3~5秒，用蒸馏水洗净后在0.1mol/L HCl溶液中浸泡复新。

（6）被测溶液中若含有易污染敏感玻璃球泡的物质，会使电极钝化而读数不准。可根据污染物的性质用适当溶液清洗使电极复新。清洗时不能用四氯化碳或四氢呋喃等溶剂。

十一、分光光度计的使用

分光光度计是根据物质对光的吸收程度进行定性或者定量分析的仪器。常用可见光分光光度计的型号有721型、722型、7200型等。这里介绍721型分光光度计。

（一）基本原理

分光光度计利用单色器获得单色光，由于单色光的谱带较宽，一般适用于有标准品对比下的定量测定。只有可见光源的分光光度计，只能用于有色溶液的比色测定。

朗伯－比尔定律（Lambert-Beer Law）是吸收光度法的基本定律，是描述物质对单色光吸收的强弱与吸光物质的浓度和厚度间关系的定律。Lambert-Beer定律的数学表达式为：

$$-\lg \frac{I}{I_0} = Ecl$$

上式中，I 为透射光强度；E 为吸光系数（absorptivity）；c 为溶液浓度；l 为溶液厚度；I/I_0 是透光率（transmittance；T），常用百分数表示；又以 A 代表 $-\lg T$，并称之为吸光度（absorbance），于是：

$$A = -\lg T = Ecl \quad 或 \quad T = 10^{-A} = 10^{-Ecl}$$

上式说明单色光通过吸光介质后，透光率 T 与浓度 c 或厚度 l 之间的关系是指数函数的关系。例如，浓度增大一倍时，透光率从 T 降至 T^2。而吸光度与浓度或厚度之间是简单的正比关系。吸光系数 E 的物理意义是吸光物质在单位浓度及单位厚度时的吸光度。在给定单色光、溶剂和温度等条件下，吸光系数是物质的特性常数，表明物质对某一特定波长光的吸收能力。不同物质对同一波长的单色光，可有不同吸光系数，吸光系数愈大，表明该物质的吸光能力愈强，测定的灵敏度愈高，所以吸光系数是定性和定量的依据。吸光系数的表示方式有如下两种。

（1）摩尔吸光系数　是指在一定波长时，溶液浓度为 1mol/L，厚度为 1cm 的吸光度，用 ε 或 E_M 标记。

（2）百分吸光系数或称比吸光系数　是指在一定波长时，溶液浓度为 1%（W/V），厚度为 1cm 的吸光度，用 $E_{1cm}^{1\%}$ 表示。

吸光系数两种表示方式之间的关系是：

$$\varepsilon = \frac{M}{10} \cdot E_{1cm}^{1\%}$$

式中 M 是吸光物质的摩尔质量。摩尔吸光系数一般不超过 10^5 数量级，通常 ε 在 $10^4 \sim 10^5$ 之间为强吸收，小于 10^2 为弱吸收，介于两者之间称中强吸收，吸光系数 ε 或 $E_{1cm}^{1\%}$ 不能直接测得，需用已知准确浓度的稀溶液测得吸光度换算而得。例如，氯霉素（M 为 323.15）的水溶液在 278nm 处有吸收峰。设用纯品配制 100ml 含有 2.00mg 的溶液，以 1.00cm 厚的吸收池在 278nm 处测得透光率为 24.3%。则：

$$E_{1cm}^{1\%} = \frac{-\lg T}{c \cdot l} = \frac{0.614}{0.02} = 307 \quad \varepsilon = \frac{323.15}{10} \times E_{1cm}^{1\%} = 9920$$

如果溶液中同时存在两种或两种以上吸光物质（a, b, c, …）时，只要共存物质不互相影响性质，即不因共存物而改变本身的吸光系数，则总吸光度是各共存物吸光度的和，即 $A_{总} = A_a + A_b + A_c + \cdots\cdots$，而各组分的吸光度由各自的浓度与吸光系数所决定。吸光度的这种加和性是计算分光光度法测定混合组分的依据。

（二）721 型分光光度计的使用

1. 721 型分光光度计的结构特点和光学线路

（1）结构特点　721 型分光光度计（图 2-21）用体积很小的晶体管稳压电源代替了笨重的磁饱和稳压器。用真空光电管作为光电转换元件。放大器以结型场效应管作为输入极，发挥了其高输入阻抗、低噪音的优点。放大后的微电流推动指针式微安表，以此代替了体积较大而且容易损坏的灵敏悬镜式光点检流计。由于整机系统的改进，故体积减小，并且稳定性和灵敏性都有所提高。

（2）光学线路　由图 2-22 可见由光源灯发出的白光照射到聚光透镜上，会聚光再经过平面镜（反射镜）反射至入射狭缝，进入单色器内，射至准直镜，以一束平行光射向棱镜（背面镀铝），光线进入棱镜后被色散，在铝面上从原路反射回来，经准直镜反射后又会聚在出射狭缝上，经聚光透镜后，照射至比色池，未被吸收的光波通过光门至光电管产生电流。

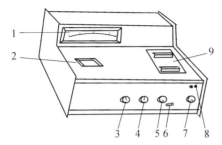 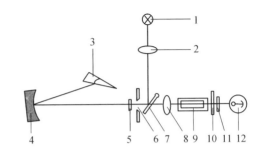

1-电表，2-波长读数盘，3-波长调节旋钮，
4-"0"透光率调节旋钮，5-"100"透光率调节旋钮，
6-比色池拉杆，7-灵敏度选择旋钮，8-电源开关，
9-比色池暗盒盖

图2-21　721型分光光度计结构示意图

1-光源灯（12V 25W），2-聚光透镜，3-色散棱镜，
4-准直镜，5-保护玻璃，6-狭缝，7-反光镜，
8-聚光透镜，9-比色池，10-光门，11-保护玻璃，12-光电管

图2-22　721型分光光度计光学线路示意图

2. 721型分光光度计的使用方法

（1）在仪器尚未接通电源时，电表的指针必须位于"0"刻线上，否则应用电表上的校正螺丝进行调节。

（2）接通电源开关（接220V交流电），打开比色槽暗箱盖，使电表指针处于"0"位，预热20分钟后，选要用的单色波长和相应的灵敏度档，用调"0"透光率调节器校正电表"0"位。

（3）合上比色槽暗箱盖，比色池处于空白校正位置，使光电管受光，旋转"100"透光率调节器，调节光电管输出的光电讯号，使电表指针正确处于100%。

（4）按上述方法连续几次调正"0"位和100%位置。

（5）把待测液置于比色池中，把待测溶液比色池处于光路中，按空白校正方法，测定，记录吸光度或透光率。

（6）测定完毕，切断电源，开关置于"关"位。洗净比色池。在比色槽暗箱中放好干燥硅胶。

3. 分光光度计的维护及注意事项

（1）将仪器安放在干燥的房间内，置于坚固平稳的工作台上，室内照明不宜太强。热天不能用电风扇直接向仪器吹风，防止灯丝发光不稳。

仪器灵敏度档的选择是根据不同的单色光波长和光能量不同分别选用，第一档为1（为常用档）、灵敏度不够时再逐级升高，但改变灵敏度后须重新校正"0"和"100%"。选用原则是能使空白档良好地用"100"透光率调节器调至100%处。

（2）在接通电源之前，应对仪器的安全性进行检查，各调节旋钮的起始位置应该正确，然后接通电源。

（3）仪器各部存放有干燥剂筒处应保持干燥，发现干燥剂变色立即更换或烘干再用。

（4）仪器长期工作或搬动后，要检查波长精度等，以确保测定结果的精确。

（5）在使用过程中应注意随时关闭好遮盖光路的闸门（打开比色池暗盒盖）以保护光电池。

（6）仪器连续使用时间不宜过长，更不允许仪器处于工作状态而测定人员离开工作岗位。最好是工作2小时左右让仪器间歇半小时左右再工作。

Chapter 2　Common Instruments and Basic Operation

1. Weighing

1.1　Direct Weighing Method

After the zero point of balance is set, put the object to be weighed directly on the weighing plate (note that the chemical reagent cannot directly contact the weighing plate), and the reading obtained is the mass of the weighed object. This weighing method is suitable for weighing clean and dry utensils, rod or block metals and other solid samples that are not easy to deliquesce or sublimate. Note that it is not allowed to take and place the weighed object directly by hand, but suitable methods such as wearing Jersey fabric gloves, padding paper strips, using tweezers or pliers, etc.

1.2　Decrement Weighing Method

This method does not require afixed value for the weight of the sample, but only within the required range. It is suitable for weighing substances that are easy to absorb water, oxidize or react with CO_2. Such substances can be weighed in a weighing bottle with a cover to prevent moisture absorption and dust prevention. The weighing sequence is as follows:

First, put a proper amount of sample in the weighing bottle (if the sample has been dried, it should be put in the desicator to cool to room temperature), cover it with clean small paper or plastic film, put it on the weighing bottle, then put the weighing bottle on the analytical balance to accurately weigh its weight, set it as W_1 (g). Take out the weighing bottle, cover it with clean small paper strip or plastic film, cover it on the weighing bottle cap, take down the weighing bottle cap above the container mouth, tap the upper part of the bottle gently with the weighing bottle cap to make the sample fall into the container slowly, as shown in Figure 2-1, then slowly erect the bottle, tap the upper part of the bottle mouth with the bottle cap to make the sample stuck in the bottle fall into the bottle, and cover the bottle cap properly. Then put the weighing bottle back to the pan for weighing, and repeat the operation until the weight of the poured sample meets the requirements. If the weight of weighing bottle and sample is W_2 (g) after the first sample is poured out, the weight of the first sample is W_1-W_2 (g).

Figure 2-1　Schematic diagram of knocking out sample from weighing bottle

2. Instruments Commonly Used in Inorganic Chemistry Experiments

instruments	specifications	general purpose	precautions for use
test tube / test tube rack	Test tube: hard and soft; with graduation, without graduation; with branch, without branch, etc. In general, the diameter of nozzle×tube length (mm) is used for the graduation-free test tube, such as 10×75, 15×150, etc. Graduated test tube is classified by capacity (ml), such as 5, 10, 15ml…… Test tube rack: wood, plastic or metal	Test tube: (1) Small amount of reaction vessel, convenient for operation and observation (2) The test tube with branch can be equipped with gas generator, gas washing device and gas product inspection Test tube rack: be used to place test tube	Test tube: (1) Hard test tube can be heated to high temperature, but it is not suitable for sudden cooling. Soft test tube is easy to break when the temperature changes rapidly (2) When heating, the nozzle of the test tube should not be facing the person, and the liquid contained should be no more than 1/3 of the capacity Test tube rack: metal tube rack should be protected from acid and alkali corrosion
centrifuge tube	There are two kinds: graduated and no graduated. The specifications are in ml, such as 5, 10ml……	(1) Small amount of reaction vessel (2) Precipitation and separation	It cannot be heated directly but can be heated in water bath
beaker	Classified by capacity (ml), such as 50, 100, 500ml…… Divided into hard and soft, with graduation and without graduation, etc.	(1) Reaction vessel, easy to mix evenly (2) Preparation of solution	(1) Dry the outer wall of the beaker before heating, and pad asbestos gauze under the beaker during heating to prevent uneven heating (2) The reaction liquid shall not exceed 2/3 of the beaker capacity to avoid overflow
conical flask/Erlenmeyer flask	By capacity (ml), such as 50, 100, 250ml…… Divided into hard, soft, with plug, without plug, etc. Erlenmeyer flask with plug is also called iodine measuring bottle or iodine flask	(1) Reaction vessel, easy to swing (2) Suitable for titration operation	Do not heat directly. When heating, put asbestos gauze or water bath
drip plate/spot plate	Glazed porcelain plate, white and black	Carry out spot reaction, observe the formation of precipitation and color changed	(1) No heating (2) It cannot be used in the reaction of hydrofluoric acid and concentrated alkali solution

Continued

instruments	specifications	general purpose	precautions for use
measuring cylinder and cup	Classified by the maximum capacity (ml) that can be measured, such as 10, 50, 100ml……	Measure a certain volume of liquid	(1) cannot be used as a reaction vessel, cannot be used heating (2) When reading the graduation, make sure line of sight should be level with the liquid meniscus, and the graduation tangent to the lowest point of meniscus shall be read
wide-mouth bottle, narrow-mouth bottle	Material: glass or plastic Specification: wide-mouth, narrow-mouth; colorless, brown By capacity, e.g. 125, 250, 500, 1000ml, etc.	Wide-mouth bottle usually for solid reagent and narrow-mouth bottle for liquid reagent	(1) No heating (2) When taking the reagent, the cap should be placed on the table upside down (3) Use rubber stoppers or plastic bottles to contain alkaline substances (4) Brown bottles for light decomposable substances
medicine spoon	By size Made of porcelain, bone, plastic, metal alloy and other materials	Used for taking solid reagents	(1) The size of the spoon should be selected so that the reagent can be put into the container after being taken (2) After taking one medicine, spoon must be washed and dried before taking another medicine
mortar	Materials: porcelain, iron, glass, agate, etc. Specification: by bowl diameter (mm), such as 60, 75, 90mm	(1) Grinding solid matter (2) Two or more drugs are grinded and mixed evenly	(1) Cannot be used as reaction vessel (2) Can only be used to grind, not to knock or dry (3) Explosive materials can only be crushed gently, not ground
watch glass	Classified according to the diameter (mm), such as 45, 65, 75, 90mm	(1) A small amount of reagent reaction container (2) Used as the cover of evaporating dish, beaker, etc.	(1) Do not heat directly with fire (2) When used as a cover, its diameter should be slightly larger than that of the covered container
funnel	According to the caliber and funnel neck length, such as 6cm long neck funnel and 4cm short neck funnel	For filtering or pouring liquid	Cannot be used to heat directly with fire

Continued

instruments	specifications	general purpose	precautions for use
Buchner funnel, filter flask/ suction flask	Buchner funnel: made of porcelain or glass. Classified according to the diameter, in cm Suction flask: in ml	Vacuum filtration	(1) The filter paper should be slightly smaller than the inner diameter of the funnel, and the holes at the bottom should be covered to avoid leakage (2) Pump first, and then filter. When the filtration is stopped, vent first, and then turn off the vacuum pump (3) No direct heating
evaporating dish	Materials: porcelain, quartz and platinum products Classified by the diameter (cm) of the upper mouth or capacity (ml). It can be divided into flat bottom and round bottom	(1) Evaporation concentration (2) Reaction vessel (3) Scorching solid	(1) High temperature resistance, direct heating (2) Be careful not to break it. Do not cool it suddenly in high temperature
volumetric flask	Classified by volume (ml), component in formula (in) and quantity out formula (ex) The colors are colorless and brown	For preparation of standard solution	(1) It cannot be heated directly or used to store solution instead of reagent bottle (2) The brown volumetric flask is used to prepare the standard solution of medicine which is easy to decompose in the visible light (3) Corks and bottles are matched and cannot be interchanged
(a) transferring pipette (b) graduated pipette	Classified by the maximum graduation (ml), such as 1, 2, 5, 10, 20ml, etc. Pipette: (a) transferring pipette/non graduated pipette (b) graduated pipette	Used to precisely measure a certain volume of liquid	(1) Do not heat and dry (2) Use a small amount of liquid to be taken for three times (3) Generally, do not blow out the last drop of liquid left in the pipette, except for the full flow pipette with the word "blow"

Common Instruments and Basic Operation   Chapter 2

Continued

instruments	specifications	general purpose	precautions for use
weighing bottle	Height (a) and flatness (b) are classified by outer diameter (mm) ×height (mm)	For accurately weighing a certain amount of solid	(1) The bottle and lid are matched and cannot be mixed. (2) Do not heat with fire (3) It should be cleaned when it is not used, and pad a small piece of paper at the grinding mouth to prevent adhesion
desiccator	Made of glass. It can be divided into ordinary desiccator and vacuum desiccator; colorless and brown	(1) Dry and store chemicals with desiccants at the bottom (2) In quantitative analysis, put the burnt crucible in it for cooling (hot chemicals should be cooled before being put into the desiccator)	(1) The burned items should be slightly cold before being put into the desiccator (2) The desiccant shall be replaced in time, and vaseline shall be evenly applied at the grinding mouth of the desiccator cover (3) Brown desiccator should be used for the sample easy to decompose in light
crucible	Materials: porcelain, quartz, zirconia, iron, nickel, platinum, etc. Classified by capacity (ml), and 30ml is commonly used	High temperature resistant, for burning solid or heat substances vigorously	(1) It can be placed on the mud triangle and directly heated by fire (2) The hot crucible is clamped with crucible tongs and placed on the asbestos gauze (3) Hot crucible shall not be cooled suddenly to avoid splashing water
crucible tongs	Iron or copper alloy products, with chromium and nickel plating on the surface	At high temperature, take the crucible or lid	(1) When the hot crucible is clamped, the tip shall be preheated to prevent the crucible from cracking due to local sudden cooling (2) Place it with the tip facing up when it is not used. (3) Avoid contact with corrosive liquids
tripod	Iron products, can be divided by size and height	Place larger or heavier heating containers	Put the asbestos gauze first, then the heating container, excluding the water bath pot

Continued

instruments	specifications	general purpose	precautions for use
clay triangle	There are different sizes, classified according to the length of each side (cm)	Crucible and small evaporating dish for holding heating	(1) Avoid dripping cold water on the hot clay triangle to prevent the porcelain pipe from breaking (2) When selecting the mud triangle, make the upper part of the crucible on it not more than 1/3 of its height
Iron stand/iron frame platform, iron ring, flask clamp/iron clamp	A piece of iron, usually rubber or plastic The iron frame platform is classified according to the height (cm); the iron ring is classified according to the diameter (cm); the iron clamp is classified according to the size, including double clamps, three clamps and four clamps	(1) For fixed reaction vessel (2) Iron ring can be used instead of funnel frame or mud triangle support frame	(1) The iron clamp shall be raised to a proper height, and the screw shall be tightened to make it firm before the experiment (2) When the fixed instrument is on the iron frame platform, the center of gravity shall fall on the center of the iron frame platform chassis
asbestos gauze	It is made of iron wire and coated with asbestos in the middle. It can be divided according to the area and size. Its size is classified by the diameter of asbestos layer, such as 10, 15cm, etc.	When heating the glassware, it should be padded at the bottom to make it evenly heated	(1) Avoid soaking the asbestos gauze to avoid falling off of asbestos and rust damage of iron wire (2) High temperature ceramics have been used to replace asbestos because of its carcinogenesis
brush	In terms of size and application, such as tube brush and beaker brush	Used for washing common glassware	When washing the test tube, the hair at the front should be pinched and put into the test tube, so as to avoid the top of the wire poking the bottom of the test tube
bottles for cleaning/ washing bottle	Plastic products, by volume (ml), such as 500ml	For washing precipitates and containers with distilled or deionized water	(1) Not for tap water (2) Plastic washing bottle cannot be heated

instruments	specifications	general purpose	precautions for use
test tube clamp	There are wooden and metal products with the same shape	Used to hold the tube during heating	(1) Clamp on the upper end of the tube (about 2cm from the tube mouth) (2) Slide or remove the tube holder from the bottom of the tube (3) When heating, hold the long handle of the test tube rack. Do not hold the long handle and the short handle at the same time

Note: all materials used in the instrument are glass unless otherwise specified. The specifications listed are commonly used.

3. Washing and Drying of Instruments

3.1 Washing of Glassware

Glass instruments are often used in chemical experiments. Whether the instruments are clean or not has a great influence on the accuracy of the experimental results. Therefore, washing glass instrument is a necessary chemical experiment preparation work. Different cleaning methods can be selected according to the nature and degree of dirt.

3.1.1 General Washing

For the cleaning of glassware such as beakers, flasks, conical bottles, measuring cylinders, watch glass, reagent bottles, etc. that is commonly used in the laboratory, wet the instrument and brush first, then dip the brush in the cleaning powder and scrub the inner and outer walls of the instrument to remove the dirt on the glass surface, and then wash them with water. If the woder attached to the inner wall of the instrument is neither condensed into droplets nor flowing down, but evenly distributed to form a water film. It means that the instrument has been leaned.

3.1.2 Wash with Lotian

The inner wall of pipette, volumetric flask and burette with precise graduation should not be brushed with brush or washed with strong alkaline solvent to avoid damaging the inner wall of the volumetric glassware and affecting the accuracy. Usually use the water solution containing 0.5% synthetic detergent to soak or pour it into the volumetric glassware and shake it for a few minutes, then discard it, and then wash it with tap water. Some dirt that is difficult to clean can be washed with appropriate washing solution according to its nature. When washing with corrosive washing solution, it cannot be washed with brush. If it is acid dirt, it can be washed with alkaline detergent; otherwise, alkaline dirt can be removed with acid detergent. The oxidizing dirt can be washed with the reducing lotion, and the reducing dirt can be washed with the oxidizing lotion.

If the dirt is inorganic, it is generally washed with chromic acid lotion. Chromic acid lotion is prepared by mixing concentrated sulfuric acid and saturated potassium dichromate solution. It is mainly used for washing instruments such as burette, pipette, volumetric flask and instruments with special shapes in quantitative experiments. If the dirt is organic, $KMnO_4$ alkaline detergent is generally used for washing. The lotion is highly corrosive. If you accidentally sprinkle the lotion on clothes, skin or desktop, you should wash it with water immediately. The used lotion should be poured back to the original bottle, which can be used repeatedly. After repeated use, the chromic acid lotion will turn green (Cr^{3+}); $KMnO_4$ lotion will turn light red or colorless (Mn^{2+}), and sometimes MnO_2 precipitation will appear at the bottom. At this time, the lotion has no strong oxidizability and can no longer be used. The wash lotion shall be poured into the waste tank after treatment, and shall not be poured into the tap water tank to avoid corrosion of the sewer or pollution of the environment.

3.1.3 Washing of Special Dirt

Some instruments often deposit some dirt with known chemical composition. At this time, it is necessary to select appropriate reagent according to the nature of the dirt and remove it through chemical reaction. For example, AgCl precipitation can be washed with ammonia water, and the silver deposited at the bottom of the tube can be removed with dilute nitric acid during the silver mirror reaction experiment.

If it is not very necessary, it is not suitable to use all kinds of chemical kits and organic solvents blindly to clean the instrument, which not only causes waste but also may bring danger.

In addition to the above cleaning methods, there are also ultrasonic cleaning methods. Put the instrument to be cleaned in the container with detergent, connect the power supply and use the ultrasonic vibration to achieve the purpose of washing.

In daily experiments, we should form the habit of cleaning the glassware immediately after use, because the nature of the dirt is clear at that time, and it is easy to wash it with appropriate methods. If it is kept for a long time, it will increase the difficulty of washing.

The way to check whether the glassware is cleaned is to add water upside down. The water flows down the wall of the glassware, and the inner wall is evenly wetted with a thin water film without water drops hanging, otherwise, it needs to be washed again, so the cleaned glassware can be used for general chemical experiments. If the vessel used for refining or organic analysis is washed, in addition to the above treatment, it must be washed with distilled water or deionized water to remove the impurities introduced by tap water. The method of washing with distilled water should adopt the principle of "a small amount and many times". Generally, spray distilled water evenly on the inner wall of the instrument with a washing bottle, turn the instrument continuously, and then pour out the water. Repeat 2~3 times.

3.2 Drying of Glassware

Some instruments can be used after cleaning, but some instruments for chemical experiments must be dry. For example, non-aqueous titration in volumetric analysis, glass apparatus used in some organic chemistry experiments, etc.; the drying of the apparatus is sometimes the key to the success of the experiment. There are several drying methods.

3.2.1 Dry by Airing (Air dry)

Dry the instruments that are not urgently needed by natural airing. After cleaning, place them upside down on a proper instrument rack, and let them dry naturally at room temperature.

3.2.2 Dry by Dryer (Oven)

Put the cleaned glassware into the electric constant temperature drying oven for drying. Drain the water before putting it into the oven. If there is no water drop, put the instrument mouth upward and put it into the oven from top to bottom. Adjust the oven temperature to 105~110°C and bake for about 1 hour. For the instrument with ground glass plug, the glass plug must be taken out when drying, and the rubber products attached to the glass instrument shall also be taken out before putting into the oven. Sometimes the instrument can also be placed in a tray.

When the oven is working, do not put wet utensils into the upper layer to prevent water drops from falling and the hot utensils from breaking due to sudden cooling. After drying, the instrument can only be taken out after the temperature in the oven decreases. When taking out the glass instrument, pay attention to prevent scalding. Do not take out the very hot glass instrument and directly contact with cold water, porcelain plate and other low-temperature worktops or cold metal surfaces, so as to avoid sudden cooling and cracking.

3.2.3　Dry by the Blower / Blow Dry

Pour all the water in the cleaned glassware and put it on the air dryer to dry it with cold or hot air, or with a blower.

3.2.4　Dry with Organic Solvent

This method is suitable for the situation that the instrument needs to be dried immediately after washing. Drain the water in the cleaned glassware as much as possible, add a small amount of 95% ethanol to shake it and pour it out, shake it with a small amount of acetone (shake it with ether if necessary), and then blow it with an electric blower cold air for 1~2 minutes (the organic solvent vapor is easy to burn and explode, so you should blow the cold air first). After most of the solvent volatilizes, blow it with hot air until it is completely dry, and then use the cold air to blow off the residual steam to avoid condensation in the container and allow the instrument to cool down gradually.

It should be noted that the measuring instrument with scale cannot be dried by heating to avoid affecting the precision of the instrument. Volatile, flammable and corrosive substances shall not be put into the oven. Instruments washed with ethanol and acetone shall not be put into the oven to avoid explosion. When the grinded glass instrument and the instrument with piston are cleaned and placed, small pieces of paper shall be padded on the grinded mouth (such as volumetric flask, acid burette, etc.) to prevent it from being stuck and not easy to open after long-term placement.

4. Classification, Management and Use of Water and Chemical Reagents

4.1　Classification of Chemical Reagents

Chemical reagent is a general term for a large class of fine chemicals, which is necessary medicament for industries such as industry, agriculture, medical and health care, scientific research and teaching. There are many kinds of chemical reagents, which can be divided into many types from different angles. According to the national standard "Classification of Chemical Reagents (GB/T 37885—2019)", chemical reagents can be divided into ten categories according to their uses.

(1) Basic inorganic chemical reagent.

(2) Basic organic chemical reagent.

(3) High purity chemical reagent.

(4) Reference material/standard sample and reference substance (excluding biochemical reference material/standard sample and reference substance).

(5) Chemical reagents for chemical analysis.

(6) Chemical reagents for instrumental analysis.

(7) Chemical reagents for Life Sciences (including biochemical reference materials/standard samples and reference substance).

(8) Isotopic chemical reagent.

(9) Special chemical reagent.

(10) Other chemical reagents.

In practical application, specific chemical reagents can be divided into superior grade purity (G.R), analytical purity (A.R), chemical purity (C.P), and experimental reagent (L.R) according to product purity and impurity content according to different national standards, product purity and inpurity contert. The label colors are respectively specified as green, red, blue, brown, or yellow. Most of the commonly used reagents have the grade standards stipulated by the state, and some of the reagents without the national standards can follow the industry standards or enterprise standards. The production costs and sales price of the four standard chemical reagents decrease in turn. Therefore, when selecting reagents, we should consider comprehensively and select appropriate reagents to reduce the cost.

Generally, in an inorganic chemistry experiment, the chemically pure reagents can meet the experimental requirements. Analytical purity is suitable for analytical work, and guaranteed reagents are required for more demanding precision analysis. With the development of science and technology, some experiments put forward higher requirements for chemical reagents, so there are some special-purpose reagents such as high-purity reagents, chromatographic pure reagents, and so on.

For chemical reagent properties, it can be divided into general reagents and hazardous reagents. According to "General Rule for Clarification and Hazard Communication of Chemicals" (GB13690—2009), hazardous reagents can be divided into physical and chemical hazards, health hazards and environmental hazards. Physical and chemical hazards mainly refer to explosives, flammables, spontaneous combustibles, oxides and peroxides. Health hazards mainly refers to acute toxicity, carcinogenicity, reproductive toxicity, respiratory toxicity, and others. Environmental hazards mainly refers to the harm to the aquatic environment, acute aquatic toxicity, and others.

4.2 Management of Chemical Reagents

There are many kinds of chemical laboratory reagents, and most of them are dangerous. The scientific and standardized management of chemical reagents is the guarantee for the smooth development of experimental work and an important part of laboratory safety management. The special reagent cabinet with ventilation, drying and far away from a heat source and fire source shall be selected to store the chemical reagent. The reagent in the cabinet shall be placed according to its nature, and the solid reagent and liquid reagent shall be stored separately. Reagents can be classified according to the alphabet or the characteristics of reagents and shall be warehoused and delivered following First-in-first-out principle according to the production date and shelf life. If conditions permit, purchase, sales and inventory software can be used for unified management.

4.2.1 Storage of General Reagents

General reagents shall be stored in a cool and ventilated place, away from a heat source and water source, and avoid a humid environment. Personnel activities shall be avoided as far as possible in the storage location to prevent drugs from falling. Such reagents mainly include inorganic salts, nonvolatile organics, low ignition point organics and so on. Although the storage conditions of general reagents are relatively low, different storage methods should be adopted according to the properties of reagents.

① Easily deteriorated reagents in the presence of light   Some reagents are easy to decompose and deteriorate under strong light, such as silver oxides and salts, high mercury and mercurous oxides, halides (except Fluorine) and some salts of their salts, bromine and iodine. These reagents shall be stored in brown glass bottles, wrapped in black paper and kept away from light in the reagent cabinet.

② Easily deteriorated reagents in the presence of heat   Most biological products are easily inactivated, decomposed and moldy at high temperatures. Some reagents are easy to react at room temperatures, such as methyl methacrylate, styrene, acrylonitrile, vinyl acetylene and other polymerizable monomers. The storage temperature should be below 10°C.

③ Easily frozen reagents The melting point of some reagents is in the range of room temperature. When the temperature is higher than the melting point or falls below the freezing point, the volume of reagents will expand or contract due to melting or solidification, which is easy to cause the explosion of reagent bottles. Such as sulfuric acid, phenol, glacial acetic acid, etc. Acetic acid, commonly known as glacial acetic acid, solidifies into ice at low temperatures. Volume expansion during solidification may cause vessel rupture. Measures shall be taken to prevent bottle cracking of such reagents.

④ Easily weathered reagents When the air is too dry, the reagent containing crystal water can gradually lose crystal water and deteriorate, so that the effective component content changes. Such as crystal sodium carbonate, crystal aluminum sulfate, crystal magnesium sulfate, chalcanthite, alum and others, these reagents should be sealed and preserved.

⑤ Easily deliquescent reagents Some reagents are easy to absorb moisture in the air, which leads to deliquescence, deterioration, reduction of active ingredients and even mildew. Such as ferric chloride, anhydrous sodium acetate, methyl orange, agar, reduced iron powder, aluminum silver powder and others. The reagents which are easy to be deliquesced or deteriorated after being damped shall be stored in a desiccator.

4.2.2 Storage of Dangerous Chemicals

① Flammable

(i) Flammable liquid: some reagents are easy to volatilize into gas and burn in case of an open fire, so the liquid with a flashpoint below 25°C is usually listed as flammable. Flammable liquids are mainly organic solvents, such as methanol, ethanol, petroleum ether, ethyl acetate, ether, acetone, carbon disulfide, benzene, toluene and others, which should be stored separately and kept cool and ventilated, especially away from the fire source.

(ii) Flammable solid: sulfur, red phosphorus, magnesium powder and aluminum powder in inorganic substances, with low ignition point, belong to flammable solid. The storage place shall be ventilated and dry. White phosphorus can spontaneously ignite in the air, so it should be stored in water and placed in a dark and cool place. Check the water quantity in the bottle regularly to prevent water evaporation from exposing white phosphorus to the water surface.

② Highly toxic Highly toxic refers to reagents that invade digestive tract and can cause poisoning and death in small quantities. Such as cyanide, arsenic trioxide and another arsenide, mercuric chloride, dimethyl sulfate, soluble barium salt, lead salt, antimony salt and some alkaloids. Highly toxic reagents shall be locked in a special fixed drug cabinet, kept by a special person, and a system for the treatment of waste liquid consumption shall be established. Such substances shall not be used in case of skin wounds.

③ Strong corrosive The liquid and solid with strong corrosivity to the skin, mucous membrane, eyes, respiratory tract and articles are classified as strong corrosives. For example, strong acid, strong base, liquid bromine, phosphorus trichloride, phosphorus pentoxide, anhydrous aluminum trichloride, ammonia water, sodium sulfide, phenol, hydrazine hydrate, and others. This kind of drug should be stored in corrosion-resistant materials and isolated from other drugs.

④ Explosive Metal potassium, sodium, calcium, calcium carbide and zinc powder react violently with water and emit combustible gas, which is very easy to cause an explosion. Potassium and sodium shall be kept in kerosene, and calcium carbide and zinc powder shall be placed in a dry place. Trinitrotoluene, trinitrobenzene, azide or diazo compounds should be handled with care. The storage of these substances shall be separated from inflammable, strong oxidants, and others. When it is not used for a long time, it shall be sealed and stored, and the container containing these substances shall be placed in a groove made of cement or brick.

⑤ Strong oxidants　Strong oxidants include peroxides and their salts (hydrogen peroxide, sodium peroxide, barium peroxide), strong oxidizing oxyacids and strong oxidizing oxysalts (nitrates, chlorates, dichromates, permanganates). When heated, impacted or mixed with reducing substances, an explosion may be caused. It is required to store it in a cool and ventilated place with the maximum temperature not exceeding 30 ℃, and it shall be isolated from flammables, combustibles or oxides and others such as acids, sawdust, carbon powder, sulfide, carbohydrate, etc., and pay attention to heat dissipation.

⑥ Radioactivity　Radioactive materials shall be stored in lead containers. Special protective equipment and relevant knowledge are required for the operation of such materials to ensure personal safety and prevent contamination and diffusion of radioactive materials.

4.3　Transferring of Chemical Reagents

The name, purity and validity period of the chemical reagent shall be specified before use, and pollution label shall be avoided during use. During the operation, the cap of the reagent bottle is placed on the experimental platform backwards to prevent the cap from being polluted. Do not touch the chemical reagent by hand. After taking the reagent, cover the original bottle cap back to the original bottle and put the bottle back to its original place.

4.3.1　Transferring of Solid Reagents

① Take the solid reagent with a clean and dry spoon, which is specially used. The used spoon can only be used again after cleaning and drying.

② When take the reagent, if the amount exceeds the required amount, the extra reagent cannot be put back to the original bottle to avoid pollution.

③ When solid reagent with accurate mass is to be weighed, weighing paper should be placed on the balance. Clear it to zero and then weigh reagent. Corrosive or deliquescent reagents can be weighed in glassware.

④ When adding solid reagents to the test tube, use a spoon or put the medicine taken out on a folded weighing paper and extend it to 2/3 of the test tube. When adding lumpy solid reagents to the test tube, the test tube shall be tilted so that the reagent slides slowly into the bottom of the test tube along the inner wall of the test tube (Fig. 2-2).

Figure 2-2　Schematic diagram of transferring solid reagents

⑤ The solid with larger particle size can be measured after grinding in dry mortar.

4.3.2　Transferring of Liquid Reagents

① When take liquid reagent, a clean and dry dropper is required. When the dropper is used to take liquid medicine, the reagent shall be sucked several times with the dropper to make the reagent fully contact with the inner wall of the dropper, so as to prevent the reagent from dropping due to thermal expansion. The dropper shall not be extended into the used experimental container to avoid pollution caused by contacting the inner wall of the container (Fig. 2-3). The dropper containing medicine shall not be placed horizontally or the nozzle shall not be inclined upward to avoid liquid dropping into the rubber head. The rubber head dropper of the reagent bottle shall be dedicated to the special bottle and shall not be used in a cross way.

② For the narrow-mouth bottle that cannot use the rubber head dropper to take the reagent, the pouring method can be used to take the reagent. Remove the bottle stopper and put it back on the table. With the label of the reagent bottle facing the palm, tilt the reagent bottle, and pour the reagent into the test tube along the wall of the test tube, or pour the reagent into the beaker or the volumetric flask along the glass rod (Fig. 2-4).

③ In the property experiment, it is not necessary to measure the liquid reagent accurately. At this time, the reagent can be taken approximately by the dropper. 15~20 drops are about 1ml.

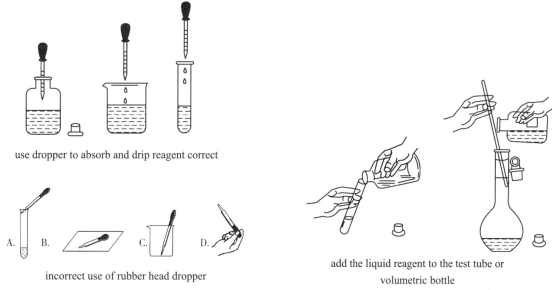

use dropper to absorb and drip reagent correct

incorrect use of rubber head dropper

Figure 2-3 Schematic diagram of rubber head dropper

add the liquid reagent to the test tube or volumetric bottle

Figure 2-4 Schematic diagram of liquid access

5. Use of Basic Measuring Instruments

5.1 Measuring Cylinder

The measuring cylinder is the most common measuring instrument in a chemical laboratory. The measuring cylinder is of various specifications, and the measuring cylinder with an appropriate range shall be selected as required. Generally speaking, the larger the measuring range, the larger the cylinder error is. Therefore, when selecting the measuring cylinder, the measuring range of the cylinder should be as close to the target volume as possible to reduce the error. For example, if a 6mL liquid needs to be taken, a measuring cylinder with a measuring range of 10mL should be selected. If a measuring cylinder with a measuring range of 100mL is selected, a large error will occur. When reading, the measuring cylinder shall be placed on a horizontal tabletop, to ensure that the line of sight is on the same horizontal line with the lowest liquid level of the solution concave in the measuring cylinder, otherwise, the measurement error will be increased. Because of the expansion and contraction of heat, too high or too low temperature will affect the measurement results, so the measuring cylinder should be used at room temperature.

5.2 Burette

Burette is the most basic measuring device in titration analysis, which consists of a long glass tube and a flow control switch. The burette body is engraved with a precise graduation, and the volume of consumed solution can be obtained by the difference between before and after titration. There are two specifications of the burette for constant analysis: 50ml and 25ml. The minimum division value of the burette is 0.1ml, and the minimum count unit of the burette added with one estimate reading should be 0.01ml. Besides, there are semi micro and micro burets with measuring ranges of 10ml, 5ml, 2ml and 1ml. The minimum graduation value is 0.05ml, 0.01ml or 0.005ml.

The burette can be divided into acid burette, basic burette and acid-basic universal burette according to different devices controlling the flow rate of solution (as shown in Fig. 2-5). Acid burette with frosted glass stopcock can hold acid, oxidizing solution and salt solution. Since silicate can be formed between the glass and alkaline solution to stick the stopcock, the alkaline solution should be placed in an basic burette. A latex tube is connected at the lower end of the basic burette body, and a glass ball with a diameter slightly larger than the inner diameter of the rubber tube is arranged in the latex tube to control the flow rate of the solution. All oxidizing solutions, such as $KMnO_4$ and I_2, which can react with the latex tube, cannot be stored in the basic burette. The structure of the acid-base universal burette is similar to the acid burette. Its stopcock is made of polytetrafluoroethylene (PTFE), which has the

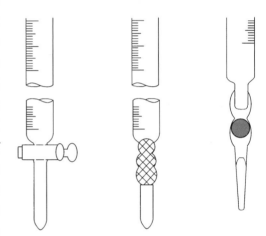

Figure 2-5 Structure diagram of burette

characteristics of acid resistance, alkali resistance and corrosion resistance. The polytetrafluoroethylene stopcock between burettes of the same batch can be used for titration of acidic, alkaline and strong oxidizing solutions.

5.2.1 Preparing

Acid burette: check whether the glass stopcock fits tightly before use, wet the stopcock with water and insert it into the stopcock, and fill the pipe with water to zero points. The tube body is fixed with a burette clamp. After 15 minutes, if the water leakage does not exceed 1 graduation (0.1ml for 50ml burette), it shall be regarded as qualified. In order to seal and lubricate the glass stopcock, it is necessary to apply vaseline to the stopcock. First, take out the stopcock, use filter paper to dry the water in the stopcock and stopcock groove, and apply a thin layer of vaseline on the thick end of the stopcock and the thin end of the stopcock groove. Carefully insert the stopcock coated with vaseline into the stopcock groove, and rotate the stopcock in the same direction until the oil film on the stopcock is even and transparent (Fig. 2-6). If it is found that the rotation of the stopcock is not flexible, or there are lines on the stopcock, it means that the oil coating is not enough; if vaseline overflows from the stopcock, or the stopcock hole is blocked, it means too much vaseline coating. After applying vaseline, use latex to cover the end of the stopcock to prevent the stopcock from falling off and damaging. Close the burette stopcock coated with vaseline, fill it with water to zero points, and fix it on the burette rack. If there is no water seepage after 2 minutes, rotate the stopcock 180° and place it for 2 minutes again. If there is no water seepage after two times, it is regarded as qualified. In case of leakage, the oil shall be applied again.

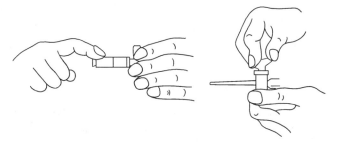

Figure 2-6 Schematic diagram of acid burette stopcock applying oil

The prepared acid burette tube shall be rinsed with tap water, soaked with chromic acid wash solution, poured out the wash solution, rinsed with tap water again to clean the residual wash solution, and then rinsed with distilled water for three times. The inner wall of the burette is a uniform water layer. If there is no water drop, it is regarded as qualified for cleaning. When cleaning, first close the stopcock, fill a small amount of detergent, hold the burette with both hands, and tilt the burette while rotating, one end of the tube body is slightly lower, one end of the stopcock is slightly higher, let the detergent wet the whole tube body. Leave a little detergent, open the stopcock to let the detergent flow out from the end of the stopcock, so that the whole burette can be fully washed.

Basic burette: select appropriate spigot, glass bead and a latex tube. The qualified basic burette shall not leak under a static state. The dripping shall be smooth during the titration, otherwise it needs to be reassembled. The cleaning method and requirements of the basic burette are the same as those of the acid burette.

Acid-basic universal burette: its leakage test method is the same as the acid burette. If there is leakage, it can be solved by adjusting the tightness of the nut of PTFE stopcock, otherwise, the burette needs to be replaced.

5.2.2 Liquid Loading

Before loading the operation solution, shake the operation solution in the storage bottle, and then rinse the burette with the operation solution for 2~3 times, each time about 10ml. After that, fill the operating fluid directly into the burette from the liquid storage bottle (do not use any other utensils, such as a pipette, beaker, funnel, to avoid contamination or dilution of the operating fluid), and finally, make the liquid level in the tube slightly higher than the maximum graduation line. Make sure that there are no bubbles in the tip of the burette filled with operating fluid. If there are bubbles, they must be eliminated. For acid burette, hold the tube body with the right hand at an angle of about 30°, and open the stopcock with the left hand quickly to make the operating fluid flow out quickly and take away bubbles. For the basic burette, it is necessary to bend the latex tube upward and squeeze the glass ball to make the operating fluid rush out quickly and eliminate the bubbles (Fig. 2-7). For acid-base universal burette, the method of exhaust bubble is the same as that of the acid burette. After filling, adjust the liquid level to zero marks.

Figure 2-7 Removing bubbles in the rubber tube of basic burette

5.2.3 Reading

Clamp the burette filled with solution vertically on the burette clamp. Due to the surface tension of the solution, the liquid level in the burette is in the shape of a crescent moon. The meniscus of the colorless solution is clear, but that of the colorless solution is not clear. Therefore, the reading methods of the two cases are slightly different. In order to read correctly, the following principles should be followed.

When reading, the burette should be vertical to the ground. After the solution is injected or discharged, wait for 1~2 minutes before reading. For colorless or light colored solutions, read the lowest point of the solid line at the lower edge of the meniscus. For this purpose, the line of sight shall be read at the same level as the lowest point of the solid line at the lower edge of the meniscus. For colored solutions, such as $KMnO_4$, I_2, and others, the line of sight should be tangent to the highest point on both sides of the liquid level.

The titration should start from the zero graduation line so that the standard solution consumed in titration can be measured within a certain volume range, and the error can be reduced. At the same time, it can avoid that the standard solution is exhausted outside the graduation line, which makes it impossible to calculate the volume difference before and after titration.

To assist in reading, a reading card may be used. This method is good for beginners to practice reading. The reading card can be made of black paper or white paper coated with a rectangle in the middle. When reading, place the reading card behind the burette so that the black part is about 1mm below the meniscus, and then you can see that the reflecting layer of the meniscus turns black, and then read the lowest point of the lower edge of the black meniscus (Fig. 2-8). The reading should be accurate to 0.01ml.

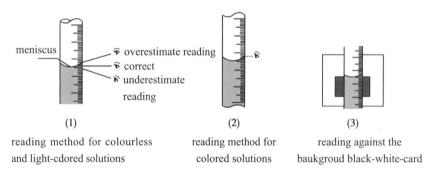

(1) reading method for colourless and light-cdored solutions

(2) reading method for colored solutions

(3) reading against the baukgroud black-white-card

Figure 2-8   Burette reading

5.2.4   Titration

When using the acid burette, the left hand should be used to control the burette stopcock. The thumb is near the side of the stopcock, and the forefinger and middle finger are far away from the side of the stopcock. Slightly bend the fingers and apply force to the left to prevent the stopcock from loosening or being pushed out. Hold the conical bottle with your right hand and shake it in the same direction. Be careful not to shake it back and forth, or the solution will spill. The initial titration speed can be slightly faster, generally 10ml/min, about 3~4 drops per second. When the standard solution enters the conical flask and causes the local indicator to change color, the color disappears slowly after the oscillation, it indicates that it is close to the end point of the titration. When approaching the end point of the titration, add one drop or half a drop, and blow a small amount of water into the washing bottle to wash the inner wall of the conical bottle, so that all the attached solution flows down, and then shake the conical bottle, so as to continue the titration until the endpoint (Fig. 2-9).

When using the basic burette, the left thumb is on the side of the burette close to the body, and the index finger is on the side of the burette far away from the body. Hold the upper part of the glass ball in the latex tube and squeeze the latex tube towards the palm to form a gap between it and the glass ball, and the solution can flow out. It should be noted that the latex tube under the glass ball cannot be pinched, otherwise it will enter the air to form bubbles. To prevent the emulsion pipe from swinging back and forth, use the middle finger and ring finger to clamp the upper part of the burette tip.

When using the acid-basic universal burette, its operation method is similar to that of the acid burette.

The titration is usually carried out in a conical flask or, if necessary, in a beaker. For the titration of iodometry and potassium bromate method, the reaction and the titration should be carried out in iodine flask. At the end of the titration, pour out the remaining solution in the burette (it can't be poured back to the original storage bottle), wash it with tap water and distilled water in turn, then fill the burette with distilled water and clamp it vertically on the burette rack. When the acid the titration tube is not used for a long time, a piece of paper shall be clamped between the grinding stopcock to prevent the adhesion of the grinding plug (Fig. 2-9).

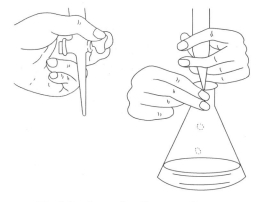

Fig. 2-9 Operation diagram of burette

5.3 Volumetric flask

The volumetric flask is a pear shaped flat bottom bottle with a narrow neck and a plug. The bottle neck is engraved with a ring marking line, indicating the volume when the liquid is filled to the marking line at the bottle body marking temperature (generally 20℃). The volumetric flask is used to prepare the accurately weighed substance into a solution of accurate concentration or to dilute the concentrated solution of known accurate concentration into a dilute solution. Generally, there are 10, 25, 50, 100, 250, 500 and 1000ml specifications.

5.3.1 Leak Detection

First, check whether the plug and the grinding mouth of the bottle body are sealed. Fill the volumetric flask with an appropriate amount of tap water, cover the plug, hold the bottom of the bottle with your right hand, stand it upside down for 2 minutes, and observe whether there is water seepage around the cork. If there is no leakage, turn the plug 180° and screw it up and turn it upside down. If there is still no leakage, it can be used. The volumetric flask and the plug should be used together. If the plug does not match the grinding mouth of the volumetric flask, it will cause liquid leakage. So the plug must be tied to the neck with string to prevent it from being lost or broken.

5.3.2 Cleaning

The volumetric flask should be soaked in chromic acid before use, and then washed with tap water and distilled water in turn. After cleaning, the inner wall without a water drop is regarded as qualified.

5.3.3 Preparation of Solution

Accurately weigh the reagent and put it into the beaker, add some distilled water to dissolve it. Transfer the solution to the volumetric flask with the aid of a glass rod. After the solution in the beaker is poured out, the beaker shall not leave the glass rod directly, but shall be lifted 1~2cm along the glass rod while the beaker is upright, and then the beaker can leave the glass rod, to avoid a drop of solution between the beaker and the glass rod flowing outside the beaker. Then wash the wall of the beaker with a small amount of distilled water for 3~4 times, and the washing solution of each time is transferred to the volumetric flask according to the same operation. When the solution reaches 2/3 of the capacity of the volumetric flask, shake the volumetric flask horizontally so that the solution is initially mixed (note that the volumetric flask cannot be reversed!) Add water to the place 2~3cm away from the mark, and use the dropper to add distilled water slowly until the lowest point of the meniscus of the solution is just tangent to the mark. Close the cork, press it with the index finger, hold the bottom of the volumetric flask with the other five fingers, turn the volumetric flask upside down, make the bubble in the bottle rise to the top, and shake it in the same direction while turning upside down. Turn and shake repeatedly until the solution in the bottle is well mixed (Fig. 2-10).

The volumetric flask should not be used for the preparation or storage of strong alkaline solution. If the solution needs to be used for a long time, the solution should be transferred to the reagent bottle for storage. The reagent bottle shall be rinsed with the solution for 2~3 times to ensure the constant concentration. The volumetric flask shall be cleaned in time after use, and shall not be baked in the oven or heated in any way. When the volumetric flask is not used for a long time, a paper strip shall be inserted between the cork and the grinding mouth of the bottle body to prevent adhesion of the grinding mouth.

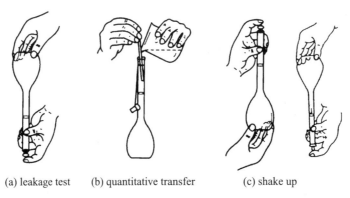

(a) leakage test   (b) quantitative transfer   (c) shake up

Figure 2-10　Use of volumetric flask

5.4　Pipettes

Pipettes are glass measuring instruments with similar structure and use, which are used to accurately transfer a certain volume of solution. The pipette is a long glass tube with a large part of the expansion in the middle (called ball part). The ball part has a thin glass neck at the top and bottom, and a mark is carved on the glass tube at the top. The commonly used specifications are 1, 2, 5, 10, 25, 50ml, and others. Compared with the pipette, the pipette has no ball and is engraved with a graduation mark on the pipette body, so it can accurately transfer the solution. Generally, the graduated pipette is slightly smaller than that of the pipette, and the commonly used specifications are 1, 2, 5, 10ml, and others.

Before use, the pipette shall be soaked with chromic acid lotion for several hours, then washed with tap water and distilled water, and finally, the water at the tip of the pipette shall be dried with filter paper. Before use, moisten the solution to be removed for 2~3 times to ensure that the concentration of the solution to be removed does not change.

When removing the solution, hold the upper end of the tube body with the right thumb and middle finger, and insert the tip into the solution. Hold the rubber suction bulb on the left hand, first squeeze out the air in the ball, and then insert the tip of the rubber suction bulb into the upper nozzle of the pipette, slowly release the rubber suction bulb, and use the atmospheric pressure to suck the solution into the pipette. When the liquid level reaches above the target graduation line, remove the rubber suction bulb, immediately press and hold the nozzle of the pipette with the index finger of the right hand, lift the tip of the pipette out of the liquid level, lean on the inner wall of the container containing the solution, slightly loosen the index finger of the right hand, and gently twist the tube body with the thumb and middle finger of the right hand to make the liquid level drop steadily. When the meniscus of the solution is tangent to the marking line, the index finger compresses the nozzle to make the liquid no longer flow out (Fig.2-11).

Take out the pipette, insert the vessel to receive the solution, and make the tip of the pipette lean against the inner wall of the vessel. Keep the pipette vertical, tilt the vessel slightly, release the right index finger, and let the solution flow down the wall naturally. After the solution completely flows out, continue to twist the pipette tip against the wall for 10~15 seconds before taking it out.

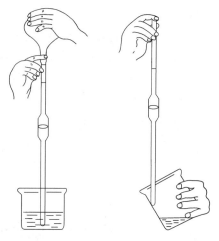

Fig. 2-11 Operation diagram of pipette (or pipette)

If the pipette is not marked with "blow", the residual solution at the end of the pipette cannot be discharged by an external force. There are some pipettes with smaller capacity, such as 0.1ml, with the word "blow" engraved on the nozzle. When using, the solution at the end needs to be blown out, otherwise, it will cause errors. In order to avoid error, the same pipette should be used as much as possible.

6. Preparing Solutions

Here we mainly introduce the preparation method of the general solution and standard solution.

6.1 Preparing General Solution

The general solution is also called auxiliary reagent test solution. When we prepare this solution, the accuracy of the concentration is not high, and generally only 1~2 significant digits are required. The preparation method of general solution mainly includes three types: direct water-soluble method, medium water-soluble method, and dilution method.

6.1.1 Direct Water-Soluble Method

This method is mainly applicable to solid reagents that are easily soluble in water without hydrolysis or with a low degree of hydrolysis, such as sodium hydroxide and sodium chloride. To prepare a solution using this method, we first weigh a certain amount of reagent using an electronic balance and transfer it to a small beaker. Then we add a small amount of distilled water to dissolve the reagent completely and dilute it with distilled water to the required volume. After mixing, we transfer the solution to the reagent bottle and label it.

6.1.2 Medium Water-Soluble Method

This method is mainly suitable for solid reagents which are easily hydrolyzed, such as ferric chloride, stannous chloride, etc. In order to avoid the hydrolysis reaction of such reagents, it is necessary to add a certain amount of medium, such as a certain concentration of acid solution or alkaline solution. The preparation process is as follows. Firstly, a certain amount of reagent is weighed using an electronic balance and transferred to a small beaker. Secondly, the acid solution or alkali solution is added to dissolve all the reagents, and then the above solution is diluted with distilled water to the required volume. Finally, after mixing, transfer the solution into the reagent bottle and label it.

6.1.3 Dilution Method

This method is mainly suitable for the solution at high concentrations, such as concentrated sulfuric acid, concentrated hydrochloric acid, concentrated nitric acid, etc. The general preparation process is as follows. Firstly, we use a graduated cylinder to take a certain volume of concentrated solution and transfer it to a small beaker. Then, we dilute it to the required volume with distilled water. After mixing, we transfer the solution to a reagent bottle and label it. However, it should be emphasized that when we use concentrated sulfuric acid to prepare dilute sulfuric acid, the concentrated sulfuric acid should be slowly added to distilled water under constant stirring, and the order of addition must not be reversed!

6.2 Preparing Standard Solution

There are two main methods of standard solution preparation: direct preparation and indirect preparation.

6.2.1 Direct Preparation Method

We weigh a certain amount of reference substance and dissolve it quantitatively. Then, we transfer it to a volumetric flask and dilute to the mark. According to the weight of the reference substance weighed and the volume of the volumetric flask, the accurate concentration of the standard solution can be calculated. This method is only applicable to reference substance, such as potassium dichromate, sodium carbonate, etc.

The preparation process is as follows. Firstly, we accurately weigh the chemicals into a clean small beaker, add a small amount of solvent, and stir with a glass rod until the chemicals is completely dissolved. Secondly, we use a glass rod to transfer the solution in the beaker into the volumetric flask. After that, we rinse the small beaker with solvent for 3~4 times, and transfer the rinse solution into the volumetric flask in the same way, thus ensuring that all chemicals are transferred into volumetric flask. Thirdly, we add the solvent to the volumetric flask until it reaches about 2/3 of the volume of the volumetric flask, and then shake the volumetric flask a few times (please do not invert the volumetric flask!). We continue to add solvent to the volumetric flask. When the liquid level is close to the mark, please let the volumetric flask stand for 1 minute. And then we use a dropper to add the solvent to the volumetric flask until the lower edge of the meniscus exactly matches the mark. Finally, we close the stopper tightly and mix the solution in the volumetric flask with several times. To obtain a solution with uniform concentration, please press the stopper with the index finger of the left hand, support the bottom of the volumetric flask with the right finger belly, and then invert and shake the volumetric flask.

6.2.2 Indirect Preparation Method

Many substances can only be prepared by the indirect method because they cannot meet the requirements of the reference substance. A certain amount of substance is roughly weighed and prepared as a solution, the concentration of which is close to the required concentration. Then, the reference substance or another standard solution with an exact concentration is used to determine the concentration of above solution. This process of using a reference substance or a standard solution with accurate concentration to determine the concentration of the solution is called calibration. For example, this method is usually used to prepare sodium hydroxide solution.

7. Using of Filter Paper and Test Paper

7.1 Using of Filter Paper

In chemical experiments, filter paper is often used as a filter medium to separate the solution from the solid. At present, commonly used filter paper is mainly divided into quantitative filter paper and qualitative filter paper. According to the pore size, each type of filter paper can be divided into three classes: fast, medium and slow. Except the quantitative analysis of precipitation, qualitative filter paper is generally used as the filter medium for the filtration experiments. The class of filter paper is determined according to the type of precipitation. Among them, filter paper with slow speed is mainly used for fine crystal precipitation; filter paper with medium speed can be used for coarse crystal precipitation; filter paper with fast speed can be used for colloidal precipitation.

The usage of filter paper for normal pressure filtration is as follows (as shown in Figure 2-12). We fold a round filter paper with appropriate size in half and unfold it to get a cone, making one side of the cone be three layers, and make the other side be one layer, and the inner angle of the cone is 60°, which just fits the inner wall of the funnel. In order to make the upper edge of the filter paper cling to the wall of the funnel without bubbles, the outermost two layers of the three-layer filter paper are often torn off a small corner. Then, we press the filter paper into the inner wall of the funnel with your index finger, moisten the filter paper with a small amount of deionized water, and gently press the filter paper around with the glass rod to remove the air bubbles between the filter paper and the wall of the funnel. Finally we add water to the edge of the filter paper in the funnel. At this time, the funnel neck should be completely filled with water to form a water column.

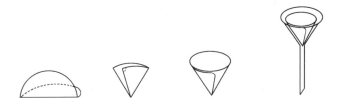

Figure 2-12 Folding and placement of filter paper

Attentions:

(1) If the inner diameter angle of the funnel is greater than or less than 60°, the angle folded by the filter paper should be appropriately increased or decreased to make the filter paper fit the wall of the funnel.

(2) The edge of the filter paper is slightly lower than the funnel mouth by 0.5~1 cm.

(3) If no water column is formed in the funnel neck, you can use your fingers to block the lower mouth of the funnel and use a wash bottle to add water between the filter paper and the funnel until the funnel neck and cone part are filled with water. After that, the air bubbles are completely discharged. Finally, please lightly press the edge of the filter paper and loosen the finger, then forming a water column.

7.2 Using of Test Paper

In chemical experiments, test paper is often used to quickly and qualitatively check the acidity and alkalinity of some solutions or gases, and the presence of certain substances. At present, there are many types of test papers, including litmus test paper, pH test paper, lead acetate test paper, starch-potassium iodide test paper, etc. The applicable scope and use methods of various test papers are as follows.

7.2.1 Litmus Test Paper

Litmus test paper is mainly used to qualitatively check the acidity and alkalinity of solutions or gases. According to the nature of the test, the test paper is mainly divided into two categories: blue litmus test paper and red litmus test paper. Among them, the blue litmus test paper is used to detect the solution or acid gas with pH <5. Red litmus test paper is used to detect solutions or alkaline gases with pH> 8.

When using litmus paper to detect the acidity and alkalinity of the solution, we first cut the paper into small pieces, and use tweezers to place a small piece of paper on a clean and dry watch glass or white drip plate. Then, we use a glass rod to dip the solution, place it in the middle of the test paper and observe the color change of the test paper. To detect the acidity and alkalinity of gas, the test paper should be moistened with distilled water, and then the test paper should be placed at the gas outlet with tweezers to observe the color change of the test paper.

7.2.2 pH Test Paper

The pH test paper is mainly used to measure the pH value of the solution. The test paper turns into different colors when it encounters solutions with different acid and alkali strengths. According to the measurable pH range, the test paper is mainly divided into two categories: extensive pH indicator paper and precise pH indicator paper. Among them, the discoloration range of the extensive pH indicator paper is from 1 to 14, which can roughly measure the pH value of the solution. The color change range of the precise pH indicator paper is relatively small, and the pH value of the solution can be measured more accurately. Common color changes of precision pH test paper include 2.7~4.7, 3.8~5.4, 5.4~7.0, 6.9~8.4, 8.2~10.0, and 9.5~13.2.

The usage of pH test paper is basically the same as that of litmus test paper. The only difference is that after the color of the test paper changes, it needs to be compared with the standard color plate to determine the pH of the solution.

7.2.3 Lead Acetate Test Paper

Lead acetate test paper is mainly used for qualitative inspection of H_2S gas. The detection principle is based on the fact that lead acetate can chemically react with H_2S to form black PbS, thus making the test paper change into dark brown with metallic luster.

When using lead acetate test paper to test H_2S gas, the test paper is first moistened with deionized water, then placed at the gas outlet with tweezers, and observe the color change of the test paper.

7.2.4 Potassium Iodide-starch Test Paper

Potassium iodide-starch test paper is mainly used for qualitative detection of oxidizing gases, such as Cl_2 and Br_2. The detection principle is based on the fact that oxidizing gas can oxidize I^- into I_2. I_2 reacts with the starch. This reaction can produce a blue-violet substance. So the color of the test paper turns into blue-violet.

When using potassium iodide-starch test paper to test the oxidizing gas, we first moisten the test paper with deionized water, then place the test paper at the gas outlet with tweezers, and observe whether the color of the test paper turns into blue-violet.

7.2.5 Precautions for Use of Test Paper

(1) To prevent pollution, the test paper needs to be sealed and taken with tweezers.

(2) The test paper cannot be directly inserted into the solution.

(3) When we measure the property of gas, the test paper cannot touch the container mouth.

8. Separation and Washing of Precipitates

8.1 Separation of Precipitation and Solution

The separation methods of solution and precipitation (or crystallization) generally include decantation, centrifugation and filtration.

8.1.1 Decantation method

When the relative density of the precipitate is large or the crystalline particles are large and can settle to the bottom of the container quickly after standing, the precipitation can be separated and washed by the decantation method. Pour the clear solution on the upper part of the precipitate into another container, and then add a small amount of detergent (such as distilled water) to wash the precipitate, stir the precipitate fully, and then pour the detergent away. Repeat the operation for 2~3 times to wash the precipitate.

8.1.2 Centrifugal separation

When a small amount of solution and precipitate are separated, centrifugal separation method is used. This method is simple and fast. For example, in the test tube reaction, it is difficult to remove the precipitate on the filter paper by the general filtration method, which is not convenient for further experiments. Generally speaking, when the relative density of precipitation is less than "1", it cannot be separated by centrifugal separation. The centrifugal instrument commonly used in the laboratory is the electric centrifuge (Fig 2-13).

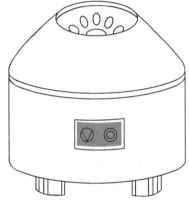

Figure 2-13   Electric centrifuge

Precautions for use of centrifuge:

(1) Put the centrifuge tube into the metal tube, the position should be symmetrical, and the weight should be balanced, otherwise the axis of the centrifuge will be damaged easily. If only one centrifuge tube needs to separate the precipitation, another empty centrifuge tube can be used to hold the corresponding mass of water, and then the centrifuge tubes are respectively symmetrically installed in the centrifuge casing to maintain the balance.

(2) Turn on the knob and gradually rotate the rheostat to make the centrifuge rotate from small to large. After a few minutes, slowly restore the rheostat to its original position and make it stop by itself.

(3) The centrifugation time and rotation speed are determined by the nature of the precipitate. The crystallized dense precipitation can be rotated at 1000 rpm and stopped after 1~2 minutes. The settling time of amorphous loose sediment is longer, and the rotating speed can be increased to 2000 rpm. If it cannot be separated after 3~4 minutes, it should try to promote the sedimentation (such as adding electrolyte or heating, etc.) and then centrifugal separation.

The operation steps of centrifugal separation are as follows.

(1) Precipitation   Add the precipitant while stirring the solution. After the reaction is complete, the solution will be centrifuged. Add another drop of reagent to the supernatant. If the supernatant is not turbid, it means that the precipitate is complete. Otherwise, add the precipitant until the precipitate is complete, and then centrifugate.

(2) Solution transfer   After centrifugation and sedimentation, separate the clear solution and

sedimentation with a pipette. The method is to squeeze the rubber head on the pipette with your fingers first to remove the air, then insert the pipette into the clear liquid gently (do not squeeze the rubber head after inserting the clear liquid), loosen the rubber head slowly, the solution will enter the pipette slowly, with the decrease of the solution in the test tube, gradually move the pipette down to all the solution is sucked into the test tube. When the tip of the pipette is close to the precipitate, take special care not to touch the precipitate (Fig. 2-14).

Figure 2-14 Separation of solution and precipitation

(3) Washing precipitates If the precipitates after centrifugation are to be dissolved before identification, the precipitates must be cleaned before dissolution. The common detergent is distilled water. After the detergent is added, stir it fully with a mixing rod, separate it by centrifugation, suck out the clear liquid with a pipette, and wash it repeatedly if necessary.

8.1.3 Filtration

Filtration is the most commonly used method to separate solution from precipitation. During filtration, the mixture of solution and precipitate passes through the filter, such as filter paper), and the precipitate is left on the filter. The solution enters the receiving container through the filter, and the resulting solution is called filtrate. The temperature and viscosity of the solution, the pressure during filtration, the pore size of the filter and the nature of the precipitate will affect the filtration speed. There are four common filtration methods in chemical experiments: atmospheric filtration, decompression filtration, thermal filtration and decantation filtration.

(1) Atmospheric filtration The method of ordinary funnel filtration under atmospheric pressure is called atmospheric pressure filtration, which is the most simple and commonly used method. The filter is a glass funnel and filter paper. When the precipitate is colloid or fine crystal, this method is better for filtration, but the disadvantage is that the filtration speed is slow. This is the simplest and commonly used filtering method. The operation steps are as follows.

Folding filter paper: take a square or circular filter paper and fold it into four layers and cut it into a sector. The circular filter paper does not need to be cut again. If the funnel specification is not standard (not 60° angle), the filter paper and the funnel are not closed. At this time, it is necessary to fold the filter paper again. It is not half folded into a proper angle, and it can be expanded into a cone with the angle of more than 60 ° after expansion, or less than 60 °. According to the angle of the funnel, make the filter paper close to the funnel, and then tear off a small angle.

Press the filter paper on the inner wall of the funnel with the index finger, moisten the filter paper with water and make it cling to the wall to remove the air bubble between the paper and the wall. When filtering, the funnel neck can be filled with filtrate. The filtrate makes the liquid in the funnel leak down with its own weight, and the filtering is greatly accelerated. Otherwise, the existence of bubbles can slow down the flow of liquid in the funnel neck and slow down the filtering speed. The edge of the filter paper in the funnel should be slightly lower than the edge of the funnel (Figure 2-15).

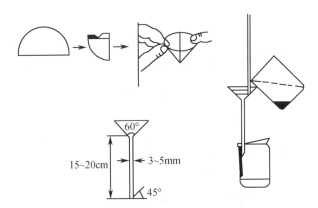

Figure 2-15 Folding method and filtering operation of filter paper

Precautions for filtration are as follows.

① The funnel is placed on the funnel holder, and the height of the funnel holder is adjusted to make the outlet of the funnel lean against the inner wall of the receiving container, so as to make the solution flow down the container wall, reduce air resistance, accelerate the filtration process, and prevent the filtrate from splashing out.

② When transferring the solution to the funnel, the decantation method should be used. Pour the solution first, and then transfer the precipitate, so as not to slow down the filtration speed because the precipitate blocks the pores of the filter paper.

③ When transferring the solution, the glass rod shall be used for drainage, and the solution shall be slowly poured into the funnel along it. The lower end of the glass rod shall be slightly touched at the three-layer filter paper to avoid the single-layer filter paper being pierced.

④ During the filtration process, there should not be too much solution in the funnel, and the liquid level should be 3~5mm lower than the upper edge of the filter paper to prevent too much solution from flowing into the receiver along the gap between the filter paper and the inner wall of the funnel, thus losing the filtering effect of the filter paper.

(2) Decompression filtration Decompression filtration is also called suction filtration or vacuum filtration. Decompression can speed up the filtration, and also can draw the precipitate dry. However, the fine particles will form a layer of dense precipitate (filter cake) on the filter paper due to vacuum suction, which makes the solution difficult to penetrate and cannot achieve the purpose of accelerating filtration. Therefore, this method is not suitable for the precipitation with too small crystalline particles. Decompression filtration is also not suitable for colloidal precipitation, because colloidal precipitation is easy to pass through the filter paper during rapid filtration (Fig 2-16).

Pay attention to the following points during operation.

① Before filtering, check whether the funnel neck is aligned with the branch pipe of the suction bottle, the long glass pipe of the safety bottle is connected with the water pump, and the short one is connected with the suction bottle.

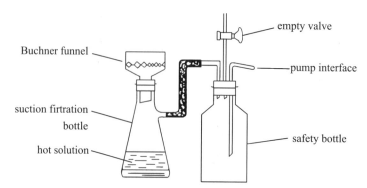

Fig 2-16 Pressure reducing filter

② The size of the filter paper should be cut to cover the magnetic hole of the funnel. First wet it with water or corresponding solvent, and then turn on the water pump to make it close to the funnel without leaving any hole. Then the filtering operation can be carried out.

③ During filtration, the upper clarifier is first injected into the funnel along the glass rod, and then the crystal or precipitation is transferred into the funnel for suction filtration. The solids that have not been completely transferred shall be washed with mother liquor and then transferred without water or corresponding solvent, so as to reduce the loss of precipitation.

④ When the filtrate will be filled with the suction bottle (but it cannot be raised to the horizontal position of the suction bottle branch pipe), pull out the rubber tube, stop the suction, take down the funnel, pour the filtrate out of the suction bottle branch pipe punching up, and then continue the suction.

⑤ In the process of suction and filtration, the water pump shall not be shut down suddenly. If it is necessary to stop suction and filtration to take out the sediment or pour out the filtrate, the rubber tube on the branch pipe of suction bottle shall be pulled out first to stop suction and filtration, and then the water pump shall be shut down, otherwise the water will be inverted.

⑥ When washing the crystal in the funnel, stop the suction filtration, let a small amount of water or corresponding solution slowly pass through the crystal, and then conduct the suction filtration and pressure drying.

Some solutions with strong acidity, strong alkalinity or strong oxidation cannot be filtered with filter paper, because the solution will damage the filter paper due to the effect of filter paper, asbestos fiber can be used instead of filter paper. This method is suitable for analysis or filtrate. There are also four common specifications of glass fused sand funnel, i.e. No. 1, No. 2, No. 3 and No. 4. No. 1 has the largest aperture. It can be selected according to different precipitated particles. But it can't be used to filter strong alkaline solution, because strong alkali will corrode glass.

(3) Thermal filtration

If the solute in the solution is easy to precipitate a large number of crystals when the temperature drops, in order not to leave the crystals on the filter paper during the filtration process, it is necessary to filter while it is hot. During the filtration, the glass funnel can be placed in the copper hot funnel (Fig. 2-17), which is filled with hot water to maintain the temperature of the solution.

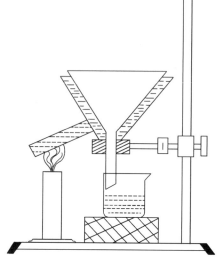

Figure 2-17 Funnel for thermal filtration

The glass funnel can also be placed on the water bath before filtration and heated with steam, and then used. This method is simple and easy. In addition, the shorter the neck of the glass funnel is, the better, so as to prevent the solution from staying too long in the funnel neck during filtering, and the blockage due to the heat dissipation and cooling and the precipitation of crystals.

(4) Decantation filtration

Before filtration, let the precipitate settle first; during filtration, do not stir the precipitate, first pour the clear liquid into the filter paper, and then transfer the precipitate to the filter paper after the clear liquid is filtered. This can prevent the sedimentation from blocking the filter hole and slowing down the filtering speed. Finally, a small amount of distilled water is blown out from the washing bottle, and the precipitate is washed 1~2 times. When it is necessary to wash the precipitate fully, it can also be washed with distilled water after pouring out the clear liquid and repeated several times.

8.2 Washing of Precipitates

The purpose of washing precipitation is to wash away the impurities adsorbed on the surface of precipitation and the mother liquor mixed in precipitation. When washing, it is necessary to minimize the dissolution loss of precipitation and avoid the formation of colloid, so it is necessary to select the appropriate washing solution. Principle of selecting washing solution: for the precipitation with small solubility and not easy to form colloid, it can be washed with distilled water; for the crystal precipitation with large solubility, it can be washed with dilute solution of precipitant, but the precipitant must be easily volatilized or decomposed and removed during drying or burning; for the precipitation with small solubility and possibly dispersed into colloid, it can be washed with dilute solution of volatile electrolyte wash.

When washing with hot washing solution, it can filter quickly and prevent the formation of colloid, but this method cannot be used for precipitation whose solubility increases rapidly with the increase of temperature.

The washing must be carried out continuously and completed once. It is not allowed to dry the precipitate for too long, especially for some amorphous precipitates, which are not easy to clean after being placed and agglomerated.

When washing the precipitate, it is necessary to wash the precipitate without increasing the dissolution loss of the precipitate. The washing effect can be improved by washing several times with a small amount of washing liquid.

9. Dissolution, Evaporation and Crystallization

9.1 Dissolution

9.1.1 Grinding of Solid

If the solid particles are large, they often need to be crushed before dissolution. In the laboratory, the grinding of solid is usually carried out in mortar. Use pestle to grind solid matter into fine particles or powder in mortar, which can accelerate solid dissolution.

Grinding operation precautions:

(1) The volume of abrasive shall not exceed 1/3 of the volume of the bowl.

(2) When grinding, use the pestle to squeeze the solid particles into the inner wall of the mortar for rotary grinding, and do not knock the solid with the pestle.

(3) Flammable, explosive and easily decomposed substances cannot be crushed by grinding.

In addition to the mortar, there are also advanced small ball mills and colloid mills in the laboratory, which can grind the diameter of solid particles to 1~5μm.

9.1.2 Dissolution

When dissolving the solid sample with solvent, the beaker shall be tilted properly before adding the solvent, and then the measuring nozzle shall be close to the wall of the beaker to let the solvent flow in slowly along the wall of the beaker; or the solvent shall flow in slowly along the glass rod through the glass rod to prevent the loss caused by the solution splashing out of the cup. After the solvent is added, stir with a glass rod to make the sample completely dissolved. For the sample that will produce gas when dissolving, first wet it with a small amount of water into paste, cover the cup with a watch glass, and then add the solvent drop by drop from the cup mouth with a dropper to prevent the generated gas from bringing out the powder sample. For the sample to be heated and dissolved, the surface dish shall be covered during heating to prevent the solution from splashing out due to violent boiling during heating. Heating and stirring can speed up the dissolution. After heating, use distilled water to wash the inner wall of the watch glass and beaker, and make the water flow down the wall of the beaker.

In the whole process of the experiment, the beaker containing the sample shall be covered with a watch glass to prevent dirt from falling. The glass rod placed in the beaker shall not be taken out at will to avoid solution loss.

9.2 Evaporation

In order to identify less ions, the solution should be concentrated before identification. The concentration of the solution is usually carried out in a small beaker. The beaker is placed in the center of the asbestos net, and the hand-held gas lamp moves back and forth below with a small fire to make the solution evaporate slowly and evenly without loss due to splashing.

If it is necessary to evaporate to dryness, the heating shall be stopped when the evaporation is near dryness, and the residual liquid shall be evaporated to dryness by itself relying on the residual heat to avoid solid splashing, and at the same time, material decomposition can be prevented.

Sometimes, if the solid left after the solution is evaporated to dryness needs strong heat burning, in this case, the evaporation of the solution should be carried out in a small crucible, and the evaporation method is the same as before. After steaming, put it on a small mud triangle and dry it with fire. The heated flame starts to be smaller, and then gradually increases the flame until it burns hot.

The evaporation concentration of the solution is usually carried out in an evaporating dish. In a few cases, it can also be heated in a beaker for evaporation and concentration, but the evaporation efficiency is poor. Attention shall be paid to the following points when evaporating the concentrated solution with an evaporating dish:

(1) The volume of liquid in the evaporator shall not exceed 2/3 of the capacity.

(2) The evaporated solution shall be carried out slowly and shall not be heated to boiling.

(3) The evaporation solution shall be carried out on the water bath (in a few cases, it can be heated on the asbestos net), and it cannot be heated directly by fire.

(4) In the process of evaporation, the solid left on the edge of liquid surface due to volume reduction shall be scraped off continuously with stirring rod.

(5) The concentration degree of the solution varies with the solubility of the solute, but the solution should not be evaporated to dryness as far as possible.

(6) Pour the liquid out of the evaporating dish from the mouth along the mixing rod.

9.3 Crystallization

All kinds of crystals have characteristic crystal shape. There are many factors that affect the crystal growth. These factors not only affect the crystal speed and crystal size, but also sometimes change the crystal shape. Therefore, in order to get a certain shape of crystal, it is necessary to have suitable crystallization conditions. Generally speaking, the crystal obtained from the thinner solution is larger and has a better crystal shape, while the crystal obtained from the thicker solution is thinner and is not easy to be complete.

9.3.1 Microcrystallization Reaction

Because all kinds of crystals have characteristic crystal shape, we can use microscope to observe the crystal shape of the reaction, and quickly make a conclusion whether some ions exist.

The operation method of microcrystallization reaction is as follows: on the dry microscope slide, each drop of test solution is about 2cm away from the reagent, and then a drop of reagent is communicated with a thin glass rod, so that the reagent reacts with the test solution slowly, resulting in crystal formation in the middle. When observing crystal shape, excessive solution should be absorbed by filter paper.

If the solution can only crystallize after concentration, the solution must be heated and evaporated on the carrier. The operation method is: first drop a drop of test solution in the center of the carrier, then use one end of the tube clamp carrier to move back and forth over the asbestos net to heat it, slowly evaporate to dryness, after cooling, add a drop of reagent to the residue, and crystal will be formed after some time.

It is necessary to use a microscope to observe the crystal formation.

The use the microscope as follows.

(1) Select the eyepiece and objective lens with appropriate magnification (the total magnification is the product of the magnification of eyepiece and objective lens).

(2) Adjust the reflector to make sure the lighting in the eyepiece is good.

(3) Place the slides on the stage. The back of the carrier should be dried to avoid contamination. The carrier shall be clamped to prevent sliding.

(4) Adjust the objective lens to be at least 5mm away from the carrier plate, and then use the left eye to look at the eyepiece and slowly raise the lens barrel until a clear image appears. If the lens barrel still fails to see the phenomenon when it rises to the highest level, the lens barrel shall be lowered to 5mm away from the carrier plate and readjusted. Never lower the lens barrel during observation to prevent the objective lens from touching the carrier plate.

(5) If the eyepiece and objective lens are polluted, they should be wiped with wiping paper instead of ordinary paper or cloth.

(6) When the microscope is not used, it should be placed in the box, and the objective lens should be placed in the special box.

9.3.2 Recrystallization

The solid compounds separated from the mixture are often impure, in which some by-products, raw materials and catalysts are often mixed. The effective way to purify these compounds is usually recrystallization with suitable solvent. The general process is as follows:

(1) The impure solid organic matter is dissolved in the solvent at or near the boiling point of the solvent to make a concentrated solution close to saturation. If the melting point of the solid organic matter is lower than the boiling point of the solvent, the saturated solution below the melting point temperature shall be made.

(2) If the solution contains colored impurities, it can be decolorized by boiling with activated carbon.

(3) Filter the hot solution to remove insoluble substances and activated carbon.

(4) The filtrate was cooled to make the crystallization precipitate from the supersaturated solution, while the impurities remained in the mother liquor.

(5) Extract and filter, separate the crystallization from the mother liquor, wash the crystallization to remove the adsorbed mother liquor. The melting point of the obtained crystal is determined after drying. If the purity of the crystal does not meet the requirements, the above operations can be repeated until the melting point does not change any more.

9.3.3 Solution Crystallization

When the filtrate is cooled rapidly and stirred violently in a cold water bath, very small crystals can be obtained. The small crystal contains less impurities, but its surface area is larger, and the impurities adsorbed on its surface are more. If you want to get even and larger crystals, you can cool the filtrate (for example, if it has crystallized in the filtrate, it can be dissolved by heating) at room temperature or under heat preservation.

Sometimes, due to the existence of tar like substances or colloidal substances in the filtrate, the crystallization is not easy to precipitate, or sometimes the formation of supersaturated solution does not precipitate crystallization. In this case, the glass rod can be used to friction the wall of the device to form a rough surface, so that the solute molecules are arranged in a directional direction and the process of forming crystallization is faster and easier than that on a smooth surface; or the crystal seed (crystal of the same substance, if there is no crystal of this substance, dip some solution with glass rod to dry slightly and then crystal will be precipitated out), supply the set crystal molecules, and make the crystal form rapidly.

Sometimes the purified substance will be separated out in the form of oil. Although the oil can be solidified after a long time of standing or enough cooling, such solid often contains more impurities (the solubility of impurities in the oil is usually higher than that in the solvent; secondly, the separated solid will also contain a part of mother liquid). The purity is not high, and it can be diluted with a large amount of solvent. Although it can prevent the formation of the oil, but It will cause a great loss of the product. At this time, the solution of oil can be heated and re dissolved, and then slowly cooled. Once the oil precipitates, the mixture will be stirred violently to make the oil solidify under the condition of even dispersion, and then the contained mother liquor will be greatly reduced. However, it is better to reselect the solvent to get crystalline products.

10. Usage of Acidometer

Acidometer (also known as pH meter) is an instrument used to measure the pH of solution, and it can also be used to measure the battery electromotive force (mV). Acidometer is mainly composed of reference electrode (saturated calomel electrode, see Figure 2-18), measuring electrode (glass electrode, see Figure 2-19) and precise potentiometer. Composite electrode is a composite made of reference electrode and measuring electrode. There are many types of pH meters commonly used in the laboratory, such as pHS-2 and pHS-3C. The structure of each type is slightly different, but the basic principles are the same.

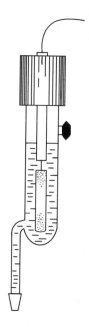

Figure 2-18　Schematic diagram of saturated calomel electrode

Figure 2-19　Schematic diagram of glass electrode

10.1　Basic Principles

In the determination of pH of solutions by direct potential method, the saturated calomel electrode is used as the reference electrode (positive electrode) and the glass electrode is used as the indicator electrode (negative electrode) which immersed in the solution to be tested, and two parts thus make up a primary battery:

(−) Ag│AgCl (s), internal fluid-filled│glass film│test solution ‖ KCl(saturated), Hg_2Cl_2 (s)│Hg (+)

The E_{MF} of the primary battery is:

$$E_{MF} = E_{calomel} - E_{glass} = K' + \frac{2.303RT}{F}\text{pH}$$

It can be seen from the above formula that the E_{MF} of the primary battery has a linear relationship with the pH of the solution, with a slope of $2.303RT/F$, which means that when the pH of the solution changes by one unit, the E_{MF} of the battery changes by $2.303RT/F$(V) (changes by about 0.059V at 25°C). In order to read the pH of the solution directly, the interval between the two adjacent readings on the pH meter is equal to the potential of $2.303RT/F$(V), which is changing with the change of temperature. Therefore, the pH meter is equipped with a temperature adjusting knob to eliminate the influence of temperature on the measurement.

In the above formula, K' is affected by many uncertain factors, so it is difficult to accurately measure or calculate it. Therefore, in the actual measurement, the "two-time measurement method" is often used. First, calibrate the pH meter with a known standard buffer solution of pH_S, which is called "positioning", so that the relationship between the battery E_S and the solution pH_S can meet the above formula, and then measure the pH_X of the solution to be measured under the same conditions, which can eliminate the influence of K', thus the pH_X of the solution to be tested is expressed as:

$$\text{pH}_X = \text{pH}_S + \frac{E_X - E_S}{0.059}　(25°C)$$

It can be seen that the pH measurement is relative, and each measurement needs to be compared with the standard buffer solution, so the accuracy of the measurement results is affected by the accuracy of the standard buffer solution pHs value.

10.2 Usage of Acidometer of pHS-3C (Figure 2-20)

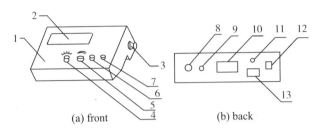

1-panel, 2-display, 3-electrode socket, 4-temperature compensation adjustent knob, 5-slope compensation adjustment knob, 6-positioning adjustment knob, 7-selection knob (pH or mV), 8-measuring electrode socket, 9-reference electrode socket, 10-nameplate, 11-fase, 12-power switch, 13-power socket

Figure 2-20 Schematic diagram of pH meters of pHS-3C

10.2.1 Start up and Install the Electrode

Insert the power cord into the power socket 13, press the power switch 12, and preheat for 30 minutes after connecting the power. Insert the electrode stem into the electrode stem socket 3, clamp the electrode clip on the electrode stem, remove the electrode sleeve at the front end of the composite electrode, and clamp the electrode on the electrode clip.

10.2.2 Calibration

The instrument should be calibrated before use. Generally, the instrument should be calibrated once a day when it is used continuously. The calibration method is as follows:

(1) Pull out the short-circuit plug at the measuring electrode socket 8 and insert the composite electrode.

(2) Turn selector knob 7 to pH.

(3) Adjust the temperature knob 4 to the temperature value of the solution to be measured.

(4) Adjust the slope knob 5 to the 100% position.

(5) Clean the electrode with distilled water, dry it with filter paper and insert it into the standard buffer solution with pH=6.86. Find out the pH of buffer solution according to the solution's temperature, and adjust the positioning knob to make the instrument display reading consistent with the pH of the buffer solution.

(6) Clean the electrode and dry it with filter paper. If the tested solution is acidic, pH = 4.00 standard buffer solution is used to adjust the slope knob to make the instrument display 4.00, if the tested solution is alkaline, pH = 9.18 standard buffer solution is used to adjust the slope knob to make the instrument display 9.18.

For the calibrated instrument, the positioning adjustment knob and slope adjustment knob shall not be turned again during measurement.

10.2.3 pH Measurement

Insert the electrode cleaned by purified water into the tested solution, shake the beaker gently to make the solution uniform, and read the pH of the solution directly on the display screen after it is still.

In general, the standard buffer solutions of pH6.86 and pH9.18 are used to calibrate the instrument when measuring the partial alkaline solution, and the standard buffer solutions of pH4.00 and pH6.86 are used when measuring the partial acid solution. During calibration, the temperature and state (static or flowing) of the standard solution shall be consistent with the temperature and state of the measured solution as much as possible. In the process of use, the instrument must be recalibrated in case of replacement of new electrode or change of "positioning" or "slope" regulator.

10.2.4 Measurement of Battery Electromotive Force (mV)

Insert the indicating electrode, reference electrode and power supply into the corresponding socket respectively, adjust the selection regulator to mV position, and put the electrode system into the tested solution for determination.

10.2.5 Preparation of Four Standard Buffer Solutions

(1) Potassium dihydrogen oxalate standard buffer solution (0.05mol/L): weigh precisely 12.71g of potassium dihydrogen oxalate [$KH_3(C_2O_4)_2 \cdot 2H_2O$] dried at (54±3)°C for 4~5h, add water to dissolve and dilute to 1000ml.

(2) Potassium hydrogen phthalate standard buffer solution (0.05mol/L): weigh precisely 10.21g of potassium hydrogen phthalate $KHC_8H_8O_4$ dried at (115±5)°C for 2~3h, add water to dissolve and dilute to 1000ml.

(3) Phosphate standard buffer solution (0.025mol/L): weigh precisely 3.55g of disodium hydrogen phosphate (Na_2HPO_4) and 3.40g of potassium dihydrogen phosphate (KH_2PO_4) dried for 2~3h at (115±5)°C, add water to dissolve and dilute to 1000ml.

(4) Borax standard buffer solution (0.01mol/L): accurately weigh 3.81g of borax $Na_2B_4O_7 \cdot 10H_2O$ (pay attention to avoid weathering), add water to dissolve and dilute to 1000ml, place it in a polyethylene plastic bottle, and keep it sealed.

Note: the water for preparing the standard buffer solution shall be pure water that has been boiled and cooled; the standard buffer solution can be kept for 2~3 months generally, but it cannot be used continuously in case of turbidity, precipitation or mildew.

10.2.6 Comparison Table of Relationship Between pH and Temperature of Standard Buffer Solution (0~50°C)

Temperature (°C)	$KH_3(C_2O_4)_2 \cdot 2H_2O$ (0.05mol/L)	$KHC_8H_8O_4$ (0.05mol/L)	$KH_2PO_4+Na_2HPO_4$ (0.025mol/L)	Borax (0.01mol/L)
0	1.666	4.003	6.984	9.464
5	1.668	3.999	6.951	9.395
10	1.670	3.998	6.923	9.332
15	1.672	3.999	6.900	9.276
20	1.675	4.002	6.881	9.225
25	1.679	4.008	6.865	9.180
30	1.683	4.015	6.853	9.139
35	1.688	4.024	6.844	9.102
38	1.691	4.030	6.840	9.081
40	1.694	4.035	6.838	9.068
45	1.700	4.047	6.834	9.038
50	1.707	4.060	6.833	9.011

10.2.7 Usage and Maintenance of Composite Electrode

(1) After removing the electrode protective sleeve, the sensitive glass bubble of the electrode shall not contact with the hard object to avoid the electrode failure due to damage or wear.

(2) The standard buffer solution with known pH must be used for calibration before measurement to ensure the reliability of the buffer solution.

(3) After the measurement, the electrode sleeve shall be put on in time, and a small amount of make-up liquid shall be put into the sleeve to keep the electrode glass bulb wet.

(4) The electrode shall not be immersed in distilled water, protein solution or acid fluoride solution for a long time, and contact with organosilicon oil shall be avoided.

(5) After long-term use of the electrode, if the slope is slightly reduced, the lower end of the electrode can be immersed in 4% HF solution for 3~5 seconds, washed with distilled water and soaked in 0.1mo/L HCl solution for renewal.

(6) If the tested solution contains substances easy to contaminate sensitive glass bubble, the electrode will be passivated and the reading will be inaccurate. According to the nature of the pollutant, the electrode can be renewed by cleaning with a proper solution. The solvent such as carbon tetrachloride or tetrahydrofuran shall not be used for cleaning.

11. Usage of Spectrophotometer

Spectrophotometer is a kind of instrument for qualitative or quantitative analysis according to the absorption degree of light. The commonly used visible spectrophotometer models are 721, 722, 7200, etc. The model 721 spectrophotometer is introduced here.

11.1 Basic Principles

Spectrophotometer uses monochromator to obtain monochromatic light. Because of the wide band of monochromatic light, it is generally suitable for quantitative determination with standard contrast. Spectrophotometer with only visible light source can only be used for colorimetric determination of colored solution.

Lambert-Beer Law（朗伯-比尔定律）is the basic law of absorption photometry, which describes the relationship between the strength of absorption of monochromatic light by matter and the concentration and thickness of absorbing matter. The mathematical expression of Lambert-Beer law is as follows:

$$-\lg \frac{I}{I_0} = Ecl$$

In the above formula, I is the transmission light intensity; E is the absorption coefficient (absorptivity, 吸光系数); c is the concentration of the solution; l is the thickness of the solution; I/I_0 is the transmittance (T; 透光率), which is commonly expressed as a percentage, and A represents $-\lg T$ which is also called the absorbance（吸光度）, therefore,

$$A = -\lg T = Ecl \quad \text{or} \quad T = 10^{-A} = 10^{-Ecl}$$

The above formula shows that the relationship between transmittance T and concentration c or thickness l is exponential function after monochromatic light passes through the absorbing medium. For example, when the concentration is doubled, the light transmittance decreases from T to T^2. The relationship between absorbance and concentration or thickness is simple proportional. The physical meaning of the absorbance coefficient E is the absorbance of the light absorbing substance at unit concentration and unit thickness. Under the conditions of given monochromatic light, solvent and temperature, the absorption coefficient is the characteristic constant of the substance, indicating the absorption ability of the substance to a certain wavelength light. Different substances can have different absorption coefficients for monochromatic light of the same wavelength. The larger the absorption coefficient is, the stronger the absorption capacity of the substance is and the higher the sensitivity of measurement is. Therefore, the absorption coefficient is the basis of qualitative and quantitative analysis. There are two ways to express the absorption coefficient:

(1) Molar absorbance coefficient（摩尔吸光系数）refers to the absorbance of solution with a concentration of 1mol/L and a thickness of 1cm at a certain wavelength, which is marked with ε or E_M.

(2) The percentage absorption coefficient or specific absorption coefficient refers to the absorbance with a solution concentration of 1% (W/V) and a thickness of 1cm at a certain wavelength, expressed in $E_{1cm}^{1\%}$.

The relationship between the two expressions of absorption coefficient is as follows:

$$\varepsilon = \frac{M}{10} \cdot E_{1cm}^{1\%}$$

M in the above formula is the molar mass of the light absorbing substance. The molar absorptivity is generally no more than 10^5 of magnitude. Generally, ε being between 10^4 and 10^5 is strong absorption, being less than 10^2 is weak absorption, and being between the two mentioned is called medium strong absorption. The absorptivity ε or $E_{1cm}^{1\%}$ cannot be measured directly, and it needs to be converted from the absorbance measured in dilute solution with known accurate concentration. For example, chloramphenicol (M=323.15) has an absorption peak at 278nm. A 100ml solution containing 2.00mg chloramphenicol is prepared with pure product, and the light transmittance is 24.3% at 278nm in a 1.00cm absorption cell. Then:

$$E_{1cm}^{1\%} = \frac{-\lg T}{c \cdot l} = \frac{0.614}{0.02} = 307 \qquad \varepsilon = \frac{323.15}{10} \times E_{1cm}^{1\%} = 9920$$

If there are two or more absorbents (a, b, c, etc.) in the solution at the same time, as long as the coexisting matter does not affect each other's properties, that is, the absorption coefficient of the coexisting matter does not change due to the coexisting matter, then the total absorbance is the sum of the absorbance of each coexistent, i.e. $A_{total} = A_a + A_b + A_c + \cdots\cdots$. The absorbance of each component is determined by its concentration and absorbance coefficient. This additivity of absorbance is the basis for calculating the determination of mixed components by spectrophotometry.

11.2 Usage of Model 721 Spectrophotometer

11.2.1 Structure Characteristics and Optical Circuit of Model 721 Spectrophotometer

(1) Structural features　The Model 721 spectrophotometer (Fig. 2-21) uses a very small transistor voltage regulator to replace the bulky magnetic saturation regulator. Vacuum photocell is used as photoelectric conversion element. The amplifier uses Junction Field Effect Transistor (JFET, 结型场效应管) as input electrode, which has the advantages of high input impedance and low noise. The amplified micro current pushes the pointer type micro ammeter to replace the large and easily damaged sensitive suspension mirror type light spot galvanometer. Due to the improvement of the whole system, the volume is reduced, and the stability and sensitivity are improved.

(2) **Optical circuit** From figure 2-22, it can be seen that the white light emitted by the light source lamp shines on the focusing lens, which will focus light and then reflect it to the incident slit through the plane mirror (reflector), enter the monochromator, and then shoot it to the collimator, and then shoot a beam of parallel light at the prism (aluminum plating on the back). After entering the prism, the light will be scattered, then it will be reflected back from the original way on the aluminum surface, and then converge on the exit slit after being reflected by the collimator. After the focusing lens is irradiated to the colorimetric cell, the unabsorbed light waves pass through the light gate to the photocell to generate current.

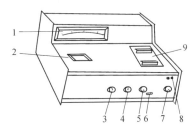

1-electricity meter, 2-wavelength reading plate, 3-wavelength adjustment knob, 4-"0" light transmittance adjustment knob, 5-"100" light transmittance adjustment knob, 6-colorimetric cell rod, 7-sensitivity selection knob, 8-power switch, 9-dark box cover of the colorimetric cell

1-light source lamp (12v 25w), 2-focusing cens, 3-dispersion prism, 4-collimator, 5-protective glass, 6-slit, 7-reflector, 8-condenser lens, 9-colorimetric cell, 10-light gate, 11-prvtective glass, 12-photocell

Figure 2-21 Structure diagram of model 721 spectrophotometer

Figure 2-22 Optical circuit diagram of model 721 spectropho tometer

11.2.2 Usage of Model 721 Spectrophotometer

(1) When the instrument is not connected to the power supply, the pointer of the meter must be on the "0" graduation, otherwise, it shall be adjusted by the correction screw on the meter.

(2) Turn on the power switch (connected to 220V AC), open the dark box cover of the colorimetric cell, make the pointer of the meter at the "0" position. After preheating for 20 minutes, select the monochromatic wavelength to be used and the corresponding sensitivity level, and use the "0" light transmittance regulator to correct the "0" position of the meter.

(3) Close the dark box cover of the colorimetric cell. The colorimetric cell is in the blank correction position, so that the photocell receives light. Rotate the "100" light transmittance regulator, adjust the photoelectric signal output by the photocell, so that the pointer of the meter is correctly at 100%.

(4) Adjust the "0" position and 100% position several times in succession according to the above method.

(5) Place the solution to be tested in the colorimetric cell, place the colorimetric cell in the optical path, measure and record the absorbance or transmittance according to the blank correction method.

(6) After the measurement, cut off the power supply and set the switch to the "off" position. Wash the colorimetric cell. Place the dry silica gel in the dark box of the colorimetric cell.

11.2.3 Maintenance and Precautions of Model 721 Spectrophotometer

(1) The instrument shall be placed in a dry room on a firm and stable working platform, and the indoor lighting shall not be too strong. In hot days, it is not allowed to blow air directly to the instrument with an electric fan, so as to prevent the unstable light of the filament.

The sensitivity level of the instrument is selected according to different monochromatic light wavelengths and different light energy. The first level is "1" (common level), and when the sensitivity is not enough, it will be increased step by step, but "0" and "100%" must be corrected after changing the sensitivity. The selection principle is to make the blank adjusted well to "100%" with "100" light transmittance regulator.

(2) Before connecting the power supply, check the safety of the instrument. The starting position of each adjusting knob should be correct, and then connect the power supply.

(3) The desiccant cylinder stored in each part of the instrument shall be kept dry. In case of discoloration of desiccant, it shall be replaced or dried for reuse.

(4) After the instrument works for a long time or it is moved away, the wavelength accuracy shall be checked to ensure the accuracy of the measurement results.

(5) In the process of use, it is necessary to close the gate that covers the light path at any time (open the dark box cover of the colorimetric cell) to protect the photocell.

(6) The continuous use time of the instrument should not be too long, and it is not allowed for the tester to leave the post while the instrument is in working state. It is better to work for about 2 hours and let the instrument rest for about half a hour before next working.

第三章　实验项目

实验一　仪器的认领和基本操作训练

微课

实验目的

1. **认领**　常用的仪器。
2. **练习**　清洗仪器。
3. 通过粗食盐的提纯，熟悉固体的取用、称量、量筒的使用、固体的加热溶解、常压过滤、减压过滤、蒸发、结晶等基本操作。

仪器、试剂及其他

1. **仪器**　台秤及砝码，量筒，滴管，玻璃棒，药匙，烧杯，研钵、杵，洗瓶，酒精灯（或煤气灯），三脚架，石棉网，玻璃漏斗，铁架，铁圈，蒸发皿，表面皿，布氏漏斗，抽滤瓶，铁夹，移液管，洗耳球，容量瓶，滴定管，锥形瓶，水浴锅，试管，试管夹，试管架，离心试管，试管刷，坩埚，泥三角，点滴板（黑、白）。
2. **试剂**　粗食盐、洗液，酒精。
3. **其他**　滤纸，火柴，蒸馏水。

实验内容

1. **认领无机化学常用仪器**　认领无机化学常用仪器，并且清点，检查有无破损。
2. **清洗仪器**

（1）对试管、烧杯、量筒等普通玻璃仪器，可在容器内先注入 1/3 左右的自来水，选用大小合适的刷子蘸取去污粉刷洗。如果用水冲洗后，仪器内壁能均匀地被水润湿而不沾附水珠，证实洗涤干净；如果有水珠沾附容器内壁，说明仍有油脂或其他垢迹污染，应重新洗涤以去除油污，必要时再用蒸馏水冲洗 2~3 次。

（2）在进行精确定量实验时，一些容量仪器的洗净程度要求较高，而且这些仪器形状又特殊，不宜用刷子刷洗，因此常先用洗液浸泡，再用自来水冲洗干净，最后用蒸馏水冲洗 2~3 次。

（3）把洗净的仪器倒置片刻，整齐地放在实验柜内，柜内铺上白纸，洗净的烧杯、蒸发皿、漏斗等倒置在纸上。试管、离心试管、小量筒等倒置在试管架上晾干。

3. **粗食盐的提纯**

（1）在粗天平的左右托盘放两张大小相同的纸，然后用药匙取 5 g 左右的粗食盐，放在左盘上。砝码放在右盘上，加减砝码至两边平衡，指针在刻度尺中间不动，盘上砝码的质量即为称量食盐的质量。

（2）将已称取的粗食盐放在研钵中磨成细匀的粉末，倾入烧杯中，用量筒量取 20ml 蒸馏水，用玻璃棒搅拌，为了加速溶解常用加热的办法。一般在三脚架上面放石棉网，然后将烧杯置于石棉网上，在网下用酒精灯加热，边搅拌边加热，直至沸腾为止。移去酒精灯，加盖表面皿，静置澄清。

（3）对澄清过的食盐溶液和不溶物进行过滤，将不溶物用少量蒸馏水洗涤 2~3 次弃去，留滤液备用；将滤液倾入干燥洁净的蒸发皿内，滤液不能超过蒸发皿容积的 2/3，以免溶液沸腾时向外飞溅。

（4）将此蒸发皿移置于三角架上，下面用酒精灯加热，当浓缩到蒸发皿底部出现结晶时，立即用玻璃棒搅拌，当快要蒸干时，应用干燥清洁的玻璃漏斗盖住，并撤去酒精灯，直至水分继续蒸干为止（为了使晶体更纯，需用重结晶法，即加少量蒸馏水溶解晶体，然后再蒸发进行结晶、分离）。

（5）用减压过滤法，得到纯净干燥的食盐晶体。减压过滤所用仪器是抽滤瓶和布氏漏斗，把食盐晶体与浓缩液转移至布氏漏斗中，进行抽滤，抽滤完毕，应先把连接抽滤瓶的橡皮管拔下，然后关闭水龙头（或关闭真空泵），以防倒吸。然后用药匙取出晶体，在台称上（精确到 0.1g）称重，计算产率。

注意事项

1. 洗液有强腐蚀性，使用时要小心。最初的洗涤废液因有酸，应倒入废物缸，不要倒进水槽。

2. 使用酒精灯时应注意以下几点。

（1）装乙醇必须在熄灯时用漏斗倒入，而且乙醇量不超过灯身容积的 2/3。

（2）点燃酒精灯时，必须用火柴，不许用 1 个已燃的酒精灯点燃另一个酒精灯，以免发生火灾或其他事故。

（3）不用时或用完后，要随时盖上灯罩，以免酒精蒸发。具体操作是盖熄后再打开片刻，然后盖上。熄灯时，千万不要用嘴吹。

（4）调节火焰，应先熄灯，用镊子夹住灯芯进行调节。灯芯不能塞得太紧。发现灯口破裂酒精灯即不能使用，以免发生火灾、爆炸。

3. 在浓缩结晶时，不能把母液蒸干（为什么）。蒸发溶液一般应在水浴锅上进行。

4. 减压过滤完毕后应先把连接抽滤瓶的橡皮管拔下，然后关闭水龙头或停真空泵，以防倒吸。

思考题

1. 如何配制铬酸洗液？
2. 怎样洗涤玻璃量器？
3. 使用洗液要注意什么？
4. 在减压过滤装置中，安全瓶的作用是什么？

Chapter 3 Experiment Items

Experiment 1 Cognition and Basic Operation Training of Instruments

Objectives

1. Cognition of glassware.
2. Practice cleaning glassware.
3. Through the purification of crude salt, familiar with the basic operations of solid extraction, weighing, cylinder use, solid heating dissolution, atmospheric pressure filtration, decompression filtration, evaporation, crystallization and so on.

Equipment, Chemicals and Others

1. Equipment

Platform scale and weight, measuring cylinder, dropper, glass rod, medicine spoon, beaker, mortar and pestle, washing bottle, alcohol burner (or gas lamp), tripod, asbestos gauze, glass funnel, iron frame, iron ring, evaporating dish, watch glass, buchner funnel, filter flask, iron clamp, transferring pipette, rubber suction bulb, volumetric flask, burette, conical bottle, water bath pot, test tube, test tube clip, test tube rack, centrifugal test tube, test tube brush, crucible, clay triangle, drip plate (black, white).

2. Chemicals

Crude salt, lotion, alcohol.

3. Others

Filter paper, matches, distilled water.

Procedures

1. Cognition of Experimental Instruments

Study inorganic chemistry commonly used instruments, and inventory, check for damage.

2. Cleaning Glassware

(1) For test tube, beaker, measuring flask and other ordinary glass instruments, the tap water of about 1/3 can be injected into the container first, and the brush of the right size can be dipped in to remove the dirt powder to brush and wash. After washing with water, if the inner wall of the instrument can be

evenly moisturized by water without adhering to the water beads, it is confirmed that the inner wall of the container is washed clean; if there is water droplets attached to the inner wall of the container, indicating that there is still grease or other mark of dirt, it should be rewashed to remove oil pollution, and rinse with distilled water for 2~3 times if necessary.

(2) In the accurate quantitative experiment, the washing degree of some capacity instruments is higher, and the shape of these instruments is special, so it is not suitable to brush them with brushes. therefore, they are often soaked in chromic acid lotion, then wash with tap water, and finally wash with distilled water for 2~3 times.

(3) Turn the washed glassware upside down for a moment, put it neatly in the experimental cabinet, put white paper in the cabinet, and invert the washed beaker, evaporated dish, funnel, etc. on the paper. The test tube, centrifugal test tube, small measuring cylinder, etc. should be put upside down on the test tube rack for drying.

3. Purification of Crude Salt

(1) Put two sheets of paper of the same size on the left and right trays of the platform scale, then use the medicine spoon to take about 5g of crude salt and place it on the left tray. Put the weight on the right tray, add and subtract the weight to balance both sides, when the pointer does not move in the middle of the scale, the mass of the weight on the right tray is that of the crud salt on the left tray.

(2) Grind the crude salt into fine powder in the mortar and pour it into the beaker. The 20ml distilled water is measured in the measuring cylinder, and the glass rod is used to stir, in order to speed up the dissolution, heating method is commonly used. Put the asbestos gauze on the tripod, then place beaker on the asbestos gauze, heat it with an alcohol burner under the gauze, stir and heat it until it boils. Remove the alcohol burner, cover the watch glass, and let it stand for clarification.

(3) Filter the clarified salt solution and insoluble matter, wash the insoluble matter with a small amount of distilled water for 2 to 3 times, and leave the filtrate for reserve; pour the filtrate into a dry and clean evaporating dish, and the filtrate should not exceed 2 to 3 of the volume of the evaporating dish, so as not to splash outward when the solution boils.

(4) Place the evaporating dish on the tripod, heat it with an alcohol burner, when crystal appears at the bottom of the evaporating dish, stir with a glass rod immediately, cover it with a dry and clean glass funnel when it is about to evaporate almost dry, remove the alcohol burner until the water continues to evaporate (in order to make the crystal purer, it is necessary to dissolve the crystal with a small amount of distilled water, and then evaporate for crystallization and separation).

(5) Pure and dry salt crystal is obtained by decompression filtration. The instrument used for decompression filtration is suction flask and buchner funnel. The salt crystal and concentrated liquid are transferred to buchner funnel for filtration. After filtering, the rubber tube connected to the suction flask should be removed first, and then the faucet (or vacuum pump) should be turned off to prevent reverse suction. And take out the crystal with the spoon, weigh with the platform scale (accurate to 0.1 g), and calculate the yield.

 Notes

1. Chromic acid lotion has strong corrosion, be careful when using. Because the original washing waste liquid has acid, it should be poured into the waste tank and not into the sink.

2. The following points should be taken into account when using alcohol burners.

① When adding alcohol, the alcohol burner must be turned off first, and then use a glass funnel to transfer the alcohol. And the amount of alcohol shall not exceed 2/3 of the volume of the alcohol burner.

② When igniting the alcohol burners, match shall be used. And a burning alcohol burner shall not be used to light another alcohol burner to avoid fire or other accidents.

③ When being not in use or after use, cover the burners at any time so as not to evaporate the alcohol. The specific operation is to put out the cover and open it for a moment, and then cover it again. Do not blow with mouth when turning off the alcohol burner.

④ When adjusting the flame, the burner should be turned off first, and then clamp the burner core with tweezers. The burner core can't be stuffed too tightly. Note that the ruptured burner mouth of alcohol burner cannot be used in order to avoid fire or explosion.

3. During the processes of concentration and crystallization, the mother liquor cannot be evaporated to dryness (why). Evaporation solution should generally be carried out on a water bath pot.

4. After decompression filtration, remove the rubber tube connected to the suction flask first, and then turn off the faucet or vacuum pump to prevent reverse suction.

Questions

(1) How to prepare the chromic acid lotion?

(2) How to wash the glass gauge?

(3) What should be paid attention while using chromic acid lotion?

(4) What is the function of the safety bottle in the decompression filter?

实验二　电解质溶液

实验目的

1. **掌握**　弱酸、弱碱的电离平衡以及平衡移动的原理；盐类水解平衡和平衡移动的原理；难溶电解质的电离平衡以及平衡移动的原理。
2. **理解**　溶度积规则的应用及测定。
3. **熟悉**　酸度计、离心机和点滴板的使用。

实验原理

1. 弱电解质的电离平衡及其平衡移动　弱电解质存在如下电离平衡，$K_a^\ominus$（$K_b^\ominus$）为弱酸（或弱碱）的电离平衡常数。$K_a^\ominus$（$K_b^\ominus$）越大，酸性（碱性）越强，影响酸碱平衡的主要因素是浓度。如果在弱电解质的电离平衡体系如醋酸中加入与弱电解质有共同离子的强电解质如醋酸钠，则平衡将向生成弱电解质醋酸的方向移动。因此同离子效应使得弱电解质的电离度降低，则醋酸的酸性下降。

$$HA \leftrightarrow H^+ + A^- \quad K_a^\ominus$$
$$BOH \leftrightarrow B^+ + OH^- \quad K_b^\ominus$$

2. 盐的水解反应　盐离子与水发生反应生成弱电解质、H^+ 和 OH^- 的反应叫作水解反应。加入水解产物可以抑制水解反应，稀释盐溶液或提高温度可以促进水解反应。

3. 难溶电解质的电离平衡及平衡移动

（1）溶度积　在难溶盐的饱和溶液中，未溶解的固体和溶解后离子之间存在下列平衡。

$$A_mB_n(s) \leftrightarrow mA^{n+} + nB^{m-}$$

其中，A_mB_n 表示难溶盐，A^{n+} 和 B^{m-} 表示 A_mB_n 溶解后产生的离子。在特定的温度下，达到平衡时 A^{n+} 和 B^{m-} 离子浓度的幂次方的乘积等于 A_mB_n 的溶度积。

$$K_{sp}^\ominus = [A^{n+}]^m \cdot [B^{m-}]^n$$

（2）同离子效应　在上述平衡中，如果增加 $[A^{n+}]$ 或 $[B^{m-}]$ 的浓度，平衡向生成沉淀 A_mB_n 的方向移动，这种现象叫作"同离子效应"。

（3）溶度积规则　根据溶度积判断沉淀的生成与溶解。当 $[A^{n+}]^m[B^{m-}]^n > K_{sp}^\ominus$，则有沉淀析出；当 $[A^{n+}]^m[B^{m-}]^n = K_{sp}^\ominus$，溶液达到饱和，但无沉淀析出；当 $[A^{n+}]^m[B^{m-}]^n < K_{sp}^\ominus$，溶液未饱和，无沉淀析出。

（4）分步沉淀　如果有两种或两种以上的离子能与沉淀剂反应生成难溶盐，则沉淀的生成先后顺序取决于所需的沉淀剂离子浓度。所需沉淀剂离子浓度小的先沉淀，所需沉淀剂离子浓度大的后沉淀。这种先后沉淀的现象称之为分步沉淀。

（5）沉淀的转化　将难溶电解质转化为另一种难溶电解质的过程称为沉淀转化。即溶解度大的难溶电解质容易转化为溶解度小的难溶电解质。

仪器、试剂及其他

固体：NH_4Cl，$NaAc$，$SbCl_3$

酸：2mol/L HAc, 6mol/L HCl, 0.1mol/L H₂C₂O₄

碱：2mol/L NH₃·H₂O, 0.2mol/L NaOH

盐：0.1mol/L CaCl₂, 1mol/L Na₂CO₃, 1mol/L FeCl₃, 0.1mol/L NaCl, 1mol/L NaCl, 1mol/L NH₄Ac, 0.1mol/L Bi(NO₃)₃, 0.1mol/L (NH₄)₂C₂O₄, 0.5mol/L NH₄SCN, 0.1mol/L AgNO₃, 0.1mol/L K₂CrO₄, 0.2mol/L MgCl₂, 0.1mol/L Pb(NO₃)₂, 0.3mol/L BaCl₂, 1mol/L NH₄Cl, 0.1mol/L KI, 0.1mol/L Na₂S

其他：酚酞指示剂、甲基橙指示剂等

实验内容

1. 弱电解质的电离平衡及平衡移动

(1) 取两支试管，每支试管中加 1ml 蒸馏水，2 滴 2mol/L 的氨水和 1 滴酚酞指示剂，摇晃均匀后观察溶液颜色。然后在一支试管中滴加少量固体 NH₄Cl，与没有加 NH₄Cl 固体的另一支试管比较颜色。解释变化的原因。

(2) 在一支试管中加入 2ml 2mol/L 的醋酸溶液，甲基橙指示剂 1 滴，观察溶液的颜色，然后再加入少量固体醋酸钠，观察溶液颜色，请从这一变化中得出结论。

(3) 取两支试管，各加 5 滴氯化镁溶液，在其中一支试管中再加 5 滴饱和氯化铵溶液，然后分别在这两支试管中加 5 滴 2mol/L 的 NH₃·H₂O 溶液，观察两支试管发生的现象有何不同。

2. 盐类的水解

(1) 取试管 4 支，分别加 5 滴 1mol/L 的碳酸钠、三氯化铁、氯化钠、醋酸铵，用 pH 试纸测定溶液的 pH。判断哪些盐类发生水解反应，写出离子方程式。

(2) 将少许固体三氯化锑（或取两滴 0.1mol/L 硝酸铋溶液）加到装有 1ml 水的试管中。有何现象发生？用 pH 试纸测定 pH。滴加 6mol/L 的盐酸使溶液刚好澄清，再加用水稀释。解释又有何现象？用平衡移动原理解释这一系列现象。

3. 难溶电解质的电离平衡及平衡的移动

（1）沉淀平衡及同离子效应　取 0.1mol/L Pb(NO₃)₂ 溶液 10 滴，加 0.5mol/L NH₄SCN 溶液，至沉淀完全。振荡试管［因为 Pb(SCN)₂ 容易发生过饱和］。可以用玻璃棒摩擦试管内壁，或剧烈地摇动试管。离心分离后，在离心溶液中加 0.1mol/L K₂CrO₄。振荡试管，有什么现象？试说明在分离出沉淀后，离心液中是否还存在 Pb^{2+}。

在试管中加入 1ml 饱和的 PbI₂ 溶液，然后加入 5 滴 0.1mol/L KI 溶液，振荡片刻。观察实验现象并解释原因。

（2）沉淀的生成　在离心管中加 2 滴 0.1mol/L Na₂S 溶液和 5 滴 0.1mol/L K₂CrO₄，加 5ml 蒸馏水稀释。再加 5 滴 0.1mol/L Pb(NO₃)₂ 溶液，观察首先生成的沉淀是黑色还是黄色。离心分离后，再向离心液中滴加 0.1mol/L Pb(NO₃)₂ 溶液，会出现什么颜色的沉淀。根据有关溶度积数据加以说明。

（3）沉淀的溶解　将 3 滴饱和草酸胺溶液与 5 滴 0.3mol/L BaCl₂ 溶液混合，产生的白色沉淀进行离心分离，弃去溶液。在沉淀中加入 6mol/L HCl 溶液。产生什么现象？写出反应方程式。

取 10 滴 0.1mol/L 的 NaCl 溶液，加 0.1mol/L 的 AgNO₃ 溶液 10 滴。离心分离，弃去溶液。在沉淀中加 2mol/L NH₃·H₂O 溶液。产生什么现象？写出反应方程式。

在试管中加 10 滴 0.2mol/L MgCl₂ 溶液，再逐滴滴加 2mol/L 氨水溶液，观察有何现象？然后再滴加 1mol/L NH₄Cl 溶液，又有何现象？写出反应式。

取两支试管各加入 1ml 0.1mol/L 氯化钙溶液，在一试管内加 5 滴 0.1mol/L 草酸铵溶液，在另一试管内加 5 滴 0.1mol/L 草酸溶液。比较两支试管中的变化。在加有 0.1mol/L 草酸铵溶液的试管内加入 5 滴 6mol/L 盐酸，会产生什么现象？再在此试管内滴加稍过量的 6mol/L 氨水，又会产生什么现象？为了判断沉淀是氢氧化钙还是草酸钙，取另一个试管，加入等量的氯化钙和氨水溶液。根据不同试管中的现象，你能判断是什么沉淀？

(4) 沉淀的转化　取 0.1mol/L Pb(NO$_3$)$_2$ 溶液 5 滴，加 3 滴 0.1mol/L NaCl 溶液。有白色沉淀生成。再加 5 滴硫代乙酰胺溶液，水浴加热。有何现象？请解释原因。

取 0.1mol/L AgNO$_3$ 溶液 10 滴，加 10 滴 0.1mol/L K$_2$CrO$_4$ 溶液中。滴加 0.1mol/L NaCl 溶液，观察有何现象？请写出反应方程式。

4. Mg(OH)$_2$ 溶度积的估算　取 50ml 烧杯 1 只，加 25ml 0.2mol/L 氯化镁溶液，并在烧杯底部垫一张黑纸。然后在氯化镁溶液中滴加 0.2mol/L 氢氧化钠溶液。不断搅拌直到沉淀形成（请直接在强光下观察）。注意氢氧化钠溶液不应过量。（为什么？）用 pH 试纸测定溶液的 pH。计算 [OH$^-$] 和 $K_{sp}^{\ominus}$。

注意事项

1. 取用液体试剂时，为避免污染试剂，严禁将试剂瓶中的滴管伸入试管内，使用完毕及时将滴管放回原试剂瓶中，不可置于实验台上。
2. 使用 pH 试纸时不得直接将试纸投入待测溶液中测试，要用洗净的玻璃棒蘸取待测溶液，滴在试纸上观察颜色的变化，并与比色卡对比。
3. 使用离心机要注意保持平衡，调整转速时不得过快。

思考题

1. 什么是同离子效应？哪个实验可以检验同离子效应的存在？
2. 沉淀平衡中的同离子效应与电离平衡中的同离子效应是否相同？
3. 沉淀生成的条件是什么？
4. 什么是分步沉淀？如何根据溶度积的计算来判断本实验中沉淀先后顺序？
5. 能否通过比较 PbCl$_2$、PbI$_2$、PbSO$_4$、PbCrO$_4$、PbS 的 $K_{sp}^{\ominus}$ 大小来解释有关沉淀转化的原因？
6. 根据平衡移动原理，试判断哪些难溶盐可以加入强酸 (如 HNO$_3$) 进行溶解？
MgCO$_3$、Ag$_3$PO$_4$、BaSO$_4$、BaSO$_3$、AgCl、MgC$_2$O$_4$

Experiment 2   Electrolyte Solution

1. Master the ionization equilibrium of weak acids, weak bases and principle of the shift of the equilibrium. Master the hydrolysis equilibrium of salts and the principle of the shift of the equilibrium.

2. Master the multiphase ionization equilibrium of some slightly soluble electrolytes and the principle of shift of the equilibrium.

3. Understand the application and test of the rule of the solubility product.

4. Familiar with how to use the pH meter, centrifugal machine, dropping board, etc.

1. The ionization equilibrium of weak electrolytes and the shift of the equilibrium

A weak electrolyte (weak acid or base) dissociates in aqueous solution as following equation. $K_a^\ominus$ ($K_b^\ominus$) stands for the dissociation extent of acid (base). The bigger the $K_a^\ominus$ ($K_b^\ominus$) value, the bigger the acid strength (base strength). The main factor that affects acid-base equilibrium is concentration. The weak electrolyte dissociates to a lesser extent when one of its ions is present in solution. For example, if sodium acetate, a strong electrolyte, is added to the acetic acid solution, the equilibrium will shift in the direction of forming acetic acid. It is called the common ion effect which makes the degree of ionization decreasing and the acidity of acetic acid decreasing.

$$HA \leftrightarrow H^+ + A^- \quad K_a^\ominus$$
$$BOH \leftrightarrow B^+ + OH^- \quad K_b^\ominus$$

2. Hydrolysis reaction of salts

The ions of salts react with water to form weak electrolytes, H^+ ion and OH^-. Such reaction is called hydrolysis reaction. Adding a product of hydrolysis can restrain the reaction, but diluting the salt solution or improving the temperature can strengthen the reaction.

3. The equilibrium of the slightly soluble electrolyte and the shift of the equilibrium

① Solubility product

In a saturated solution of some slightly soluble salts, there exists an equilibrium between the solid phase that doesn't dissolve and its ions that are formed after part of the solid dissolving. The equilibrium is as following:

$$A_mB_n(s) \leftrightarrow mA^{n+} + nB^{m-}$$

Here, A_mB_n denotes slightly soluble salt, A^{n+} and B^{m-} denote ions that are formed after part of A_mB_n dissolving. At certain temperature, the product of the powers of some relative ions is equal to the solubility product of this substance when the balance is reached.

$$K_{sp}^\ominus = [A^{n+}]^m \cdot [B^{m-}]^n$$

② Common ion effect

In the above equilibrium, if the concentration of $[A^{n+}]$ or $[B^{m-}]$ is increased, the equilibrium shifts to the formation of precipitate A_mB_n. This is called "common-ion effect".

③ The law of solubility product

Solubility product can be regarded as the standard rule to judge about the formation or the dissolution of precipitates. When $[A^{n+}]^m[B^{m-}]^n > K_{sp}^{\ominus}$, the precipitate is forming, when $[A^{n+}]^m[B^{m-}]^n = K_{sp}^{\ominus}$, the solution just turns saturated, so the precipitate is not forming yet; when $[A^{n+}]^m[B^{m-}]^n < K_{sp}^{\ominus}$, the solution is unsaturated, so no precipitate appears.

④ Fractional precipitation

If there are two or more kinds of ions which can react with a precipitating reagent to form slightly soluble salts, the sequence of the formation of precipitates is dependent on the needed concentration of the precipitating reagent ions. The lowest concentration of the ion needed means the corresponding precipitate will form first, then another precipitate will form with the greater concentration of the ion needed. We call it fractional precipitation.

⑤ Transformation of precipitate

The process of transforming a slightly soluble electrolyte to another is called precipitate transform. As a common fact, the slightly soluble electrolyte with higher value in solubility can easily be transformed to those with lower value in solubility.

Equipment, Chemicals, and Others

Solid: NH_4Cl, NaAc, $SbCl_3$

Acid: 2mol/L HAc, 6mol/L HCl, 0.1mol/L $H_2C_2O_4$

Base: 2mol/L $NH_3 \cdot H_2O$, 0.2mol/L NaOH

Salt: 0.1mol/L $CaCl_2$, 1mol/L Na_2CO_3, 1mol/L $FeCl_3$, 0.1mol/L NaCl, 1mol/L NaCl, 1mol/L NH_4Ac, 0.1mol/L $Bi(NO_3)_3$, 0.1mol/L $(NH_4)_2C_2O_4$, 0.5mol/L NH_4SCN, 0.1mol/L $AgNO_3$, 0.1mol/L K_2CrO_4, 0.2mol/L $MgCl_2$, 0.1mol/L $Pb(NO_3)_2$, 0.3mol/L $BaCl_2$, 1mol/L NH_4Cl, 0.1mol/L KI, 0.1mol/L Na_2S

Other: phenolphthalein indicator, methyl orange indicator, etc.

Experimental Procedures

1. The ionization equilibrium of weak electrolyte and its equilibrium shifting

(1) Take two test tubes, then add 1ml distilled water, 2 drops of 2mol/L $NH_3 \cdot H_2O$ and 1 drop of phenolphthalein indicator to each tube. Shake them and observe the color of the solutions. Then add a little amount of solid NH_4Cl to one tube. Please compare it with the other tube without NH_4Cl. Explain the reason of change.

(2) Add 2ml 2mol/L HAc solution to a test tube. Observe the color of the solution when 1 drop of methyl orange indicator is added. Then add a little amount of solid NaAc to it. What color appears? Please draw a conclusion from the change.

(3) Take two tubes containing 5 drops of $MgCl_2$ solution each. Add 5 drops of saturated NH_4Cl solution to one tube. Then add 5 drops of 2mol/L $NH_3 \cdot H_2O$ to each tube. Observe the different phenomena in these two tubes.

2. Hydrolysis of salts

(1) Add 5 drops of 1mol/L Na_2CO_3, $FeCl_3$, $NaCl$, NH_4Ac to each of four test tubes respectively. Determine the value of pH of the solution with pH test paper. Tell which salts are hydrolytic and write out the reaction equation.

(2) Add a little amount of solid $SbCl_3$ (or 2 drops of 0.1mol/L $Bi(NO_3)_3$ solution) to a test tube containing 1ml water. What happens? Determine its values of pH with pH test paper. Then add 6mol/L HCl until the solution just turns transparent. Dilute the solution with water. What happens then? Explain the series of phenomena with the principle of the equilibrium shift.

3. The ionization equilibrium of weak electrolyte and the shift of the equilibrium

(1) Precipitation equilibrium and common ion effect

Add 0.5mol/L NH_4SCN solution to 10 drops of 0.1mol/L $Pb(NO_3)_2$ solution until precipitate is forming completely. Shaking the test tube (Because $Pb(SCN)_2$ easily occurs over-saturated). We can rub the inner wall of the test tube with a glass rod, also we can shake the tube tempestuously. Separate the precipitate with centrifugal machine. Then add 0.1mol/L K_2CrO_4 to the above centrifugal solution. Shaking the test tuke. What happens? Try to illustrate if there exists Pb^{2+} in the centrifugal solution after separating the precipitate.

Add 1ml saturated PbI_2, solution to a test tube and add 5 drops of 0.1mol/L KI solution. Shake the test tube for a moment. What happens? Explain the reason.

(2) Formation of the precipitates

Add 2 drops of 0.1mol/L Na_2S solution and 5 drops of 0.1mol/L K_2CrO_4 to a centrifugal tube. Dilute with 5ml distilled water. Then add 5 drops of 0.1mol/L $Pb(NO_3)_2$ solution, observe the color of the precipitate first appears (black or yellow?) Separate with centrifugal machine. Then drop 0.1mol/L $Pb(NO_3)_2$ solution to the centrifugal solution above which was separated from the precipitate. What happens? (Observe the color of the precipitate). Illustrate the phenomena on the value of the solubility product.

(3) Dissolution of the precipitates

Mix 3 drops of saturated oxalate amine solution with 5 drops of 0.3mol/L $BaCl_2$ solution. White precipitate is forming. Separate with centrifugal machine and then discard the solution. Add 6mol/L HCl solution to the precipitate. What happens? Write out the reaction equation.

Mix 10 drops of 0.1mol/L NaCl solution with 10 drops of 0.1mol/L $AgNO_3$ solution. Separate the precipitate with centrifugal machine. Discard the solution. Then add 2mol/L $NH_3 \cdot H_2O$ to the precipitate. What happens? Write out the reaction equation.

Add 2mol/L $NH_3 \cdot H_2O$ drop by drop to a test tube containing 10 drops of 0.2mol/L $MgCl_2$ solution. What happens? Then with the addition of 1mol/L NH_4Cl solution, what happens? Write out the reaction equation.

Add 1ml 0.1mol/L $CaCl_2$ to two test tubes respectively. Then add 5 drops of 0.1mol/L $(NH_4)_2C_2O_4$ to one tube, and 5 drops of 0.1mol/L $H_2C_2O_4$ solution to another tube. Compare the changes in two tubes. After that, if 5 drops of HCl solution (6mol/L) is added to the first tube, what happens? Then a little excessive amount of 6mol/L $NH_3 \cdot H_2O$ is added, what happens again? In order to judge the precipitate is $Ca(OH)_2$ or CaC_2O_4, take another tube and add equal amount of $CaCl_2$ and $NH_3 \cdot H_2O$. According to the phenomena in different test tubes, can you tell what the precipitate is?

(4) Precipitate transformation

Add 3 drops of 0.1mol/L NaCl solution to 5 drops of 0.1mol/L $Pb(NO_3)_2$ solution. White precipitate is forming. Then add 5 drops of thioacetamide solution and heat it over water bath. What happens? Explain the reason.

Add 10 drops of 0.1mol/L K_2CrO_4 solution to 10 drops of 0.1mol/L $AgNO_3$ solution. Then with the addition of 0.1mol/L NaCl solution, what happens? Write out the reaction equation.

4. Predict the value of the solubility product of $Mg(OH)_2$

Add 25ml of 0.2mol/L $MgCl_2$ to a 50ml beaker which is underlaid a piece of black paper at the bottom. Then add 0.2mol/L NaOH solution to $MgCl_2$ solution. Stir constantly until precipitate is forming. (Please observe on the blazing sunshine directly). Notice that NaOH solution should not be excessive. (Why?) Determine the pH value of the solution with pH test paper. Calculate $[OH^-]$ and $K_{sp}^\ominus$.

Notes

1. When taking liquid reagent, we should not put the dropper of the reagent bottle into the test tube or on the bench in order to avoid pollution.

2. When using the pH indicator paper, we should not directly put the paper into the solution and need to dip the solution using a clean glass rod for testing, and then observe the color by comparing it with the card.

3. When using the centrifuge, we should balance the centrifugal machines and adjust the speed gently.

Questions

(1) What is common ion effect? Which experiment can test the commonion effect?

(2) Is the common ion effect in the precipitation equilibrium as same as that in acid-base ionization equilibrium?

(3) What is the condition of the formation of the precipitate?

(4) What is fractional precipitation? How to judge about the sequence of the formation of the precipitate in a certain experiment on the value of the solubility product?

(5) Can we explain the precipitate transform by comparing the $K_{sp}^\ominus$ of $PbCl_2$, PbI_2, $PbSO_4$, $PbCrO_4$, PbS?

(6) Try to judge which slightly soluble salt can be soluble in a certain strong acid (such as HNO_3) according to the principle of the equilibrium shift.

$MgCO_3$, Ag_3PO_4, $BaSO_4$, $BaSO_3$, AgCl, MgC_2O_4

实验三 缓冲溶液的配制与性质

实验目的

1. **掌握** 缓冲溶液的配制方法及其性质。
2. **熟悉** 智能酸度计的操作使用方法。
3. **了解** 使用 Excel 软件作图分析实验数据的方法。

实验原理

1. **弱电解质的电离平衡及同离子效应** 若 AB 为弱酸或弱碱，则在水溶液中存在下列平衡：

$$AB \rightleftharpoons A^+ + B^-$$

达到平衡时，各物质浓度关系满足

$$K^\ominus = \frac{C^\ominus_{A^+} C^\ominus_{B^-}}{C^\ominus_{AB}}$$

$K^\ominus$ 为标准电离平衡常数。在此平衡体系中，如加入含有相同离子的强电解质，即增加 A^+ 或 B^- 的浓度，则平衡向生成 AB 分子的方向移动，使弱电解质的电离度降低，这种现象叫做同离子效应。缓冲溶液中缓冲对的设计就依赖于此。

2. **缓冲溶液** 弱酸及其盐（如 HAc 和 NaAc）或弱碱及其盐（如 $NH_3 \cdot H_2O$ 和 NH_4Cl）的混合溶液能在一定程度上对少量外来的强酸或强碱起缓冲作用，即当外加少量酸、碱或适度稀释时，此混合溶液的 pH 变化不大，这种溶液叫作缓冲溶液。

仪器、试剂及其他

1. **仪器** 50ml 烧杯 4 个，100ml 烧杯 1 个，10ml 量筒 1 支，洗瓶 1 支，0.5ml 移液管 2 支，玻棒 1 支，洗耳球 1 个，（pHS-3C+/3C）智能酸度计 1 台。
2. **试剂** 0.2mol/L NaH_2PO_4 溶液，0.2mol/L Na_2HPO_4 溶液，1mol/L HCl 与 1mol/L NaOH 溶液。
3. **其他** 滤纸条。

实验内容

1. **（pH$_S$-3C+/3C）智能酸度计校准** 接通电源，为智能酸度计打开开关，选择显示模式为"pH"，预热 10 分钟。小心取下电极下端的保护帽，将电极玻璃球完全浸没于蒸馏水中洗净备用。

分别取适量 pH 为 4 和 9 的标准缓冲液，将测试电极下端玻璃球完全浸没于标准缓冲溶液中，待读数稳定后按机身"CAL"校准键，显示屏相应的"4"或"9"指示灯亮起，则校准成功。

2. **缓冲溶液的配制** 取 0.2mol/L NaH_2PO_4 溶液 30ml 与 0.2mol/L Na_2HPO_4 溶液 20ml 均匀混合于 50ml 小烧杯中配成缓冲比为 3:2 的 NaH_2PO_4-Na_2HPO_4 缓冲溶液 A。将缓冲溶液 A 均分至另一 50ml 小烧杯内，每杯 25ml 备用。

另取两个 50ml 小烧杯，分别装入 25ml 蒸馏水作为空白对照液 B 备用。

3. 缓冲溶液的性质测定 取盛有 25ml 缓冲溶液 A 的小烧杯，使用智能酸度计测定缓冲溶液初始 pH 并记录。分 10 次向小烧杯中加入 1mol/L HCl，每次使用移液管加入 0.5ml。搅拌均匀，待酸度计读数稳定后，测定每次加入酸后溶液 pH 并记录。

重新校准酸度计，同法操作，测定 10 次加入 1mol/L NaOH 碱液后缓冲溶液的 pH 并记录。

重新校准酸度计，同法操作，使用空白对照液 B 替换缓冲溶液 A 完成上述加酸、加碱溶液 pH 的测定工作并记录。

4. 数据分析与作图 分别将缓冲溶液甲的 20 次 pH 数据和蒸馏水的 20 次数据录入 Excel 表格中（图 3-1），以它们为数据源插入"带平滑线和数据标记的散点图"获得 pH 变化曲线，调整横纵坐标和曲线名称等得到图 3-2。

运用缓冲溶液的性质原理解释两张曲线图各自特点及其差异，完成实验报告。

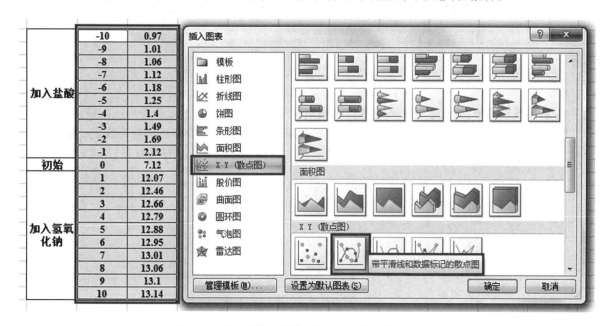

图 3-1 插入"带平滑线和数据标记的散点图"示例

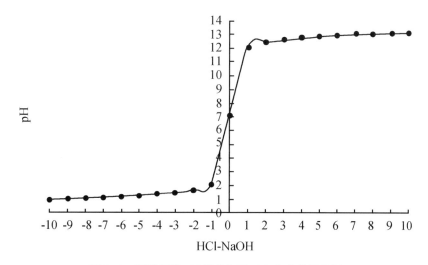

图 3-2 调整后的 pH 随外加酸碱变化曲线图示例

注意事项

1. 使用智能酸度计测定溶液 pH 时，每次测量后都要用洗瓶内的蒸馏水冲洗测定电极下端玻璃泡，并用滤纸条轻轻擦拭洗掉残留蒸馏水以使测定结果尽量准确。注意，不要刮伤电极玻璃泡，否则电极将失效。

2. 使用移液管量取外加的酸碱溶液，可固定一只量取盐酸，另一只量取氢氧化钠，合理减少因更换移液管时而造成的反复清洗润洗工作，但是第一次使用移液管时必须以蒸馏水洗涤 3 次，再用待取酸碱溶液润洗 3 次。

3. 实验完毕后，将智能电极冲洗后重新套入电极保护帽内，避免电极玻璃泡完全干燥而影响使用寿命。清理实验器具和纸屑，保持实验台面规整。

思考题

1. 对比缓冲溶液 A 与空白对照液 B 的曲线图，两者有什么差异？为什么如此不同？

2. 本实验使用缓冲比为 3∶2 的 NaH_2PO_4-Na_2HPO_4 缓冲液，实验数据所得曲线图是否沿中心对称？为什么？如果缓冲比为 2∶3，请预测实验数据所得曲线图会有怎样的变化？

Experiment 3 Preparation and Properties of Buffer Solution

1. Master the preparation method and properties of buffer solution.
2. Familiar with the method to operate the intelligent acidity meter.
3. Understand the method of using Excel software to analyze experimental data.

1. Ionization equilibrium of weak electrolyte and the common-ion effect

If AB is weak acid or weak base, the following equilibrium exists in the aqueous solution:

$$AB \rightleftharpoons A^+ + B^-$$

When the equilibrium is reached, the concentration relationship of each substance meets

$$K^\ominus = \frac{C^\ominus_{A^+} C^\ominus_{B^-}}{C^\ominus_{AB}}$$

$K^\ominus$ is the standard ionization equilibrium constant. In this equilibrium system, if a strong electrolyte containing the same ions is added, that is to say, the concentration of A^+ or B^- is increased, the equilibrium moves towards the direction of AB molecule formation, which reduces the ionization degree of weak electrolyte. This phenomenon is called the common-ion effect. The design of buffer pair in buffer solution depends on this.

2. Buffer solution

The mixed solution of weak acid and its salts (such as HAC and NaAc) or weak base and its salts (such as $NH_3 \cdot H_2O$ and NH_4Cl) can cushion a small amount of foreign strong acid or alkali to a certain extent, that is, when a small amount of acid, alkali is added or the mixed solution appropriately is diluted, the pH value of the mixed solution does not change much. This kind of solution is called buffer solution.

 Equipment, Chemicals, and Others

1. Equipment

four beakers(50ml), beaker(100ml), measuring cylinder(10ml), washing bottle, two pipettes (0.5ml), glass rod, rubber suction bulb, (pHS-3c+/3C) pH meter.

2. Chemicals

0.2mol/L NaH_2PO_4 solution, 0.2mol/L Na_2HPO_4 solution, 1mol/L HCl and 1mol/L NaOH solution

3. Others

Scraps of filter paper

Experimental Procedures

1. The calibration of (PHS-3c+/3C) pH meter

Turn on the power, turn on the switch for the intelligent acidity meter, select the display mode as "pH", and preheat for 10 minutes. Carefully remove the protective cap at the lower end of the electrode, and completely immerse the electrode glass ball in distilled water for cleaning.

Take a proper amount of standard buffer solution with pH of 4 and 9 respectively, immerse the glass ball at the lower end of the test electrode in the standard buffer solution completely, press the "CAL" calibration key of the machine body after the reading is stable, and the corresponding "4" or "9" indicator light of the display screen lights up, then the calibration is successful.

2. Preparation of buffer solution

Take 30ml of 0.2mol/L NaH_2PO_4 solution and 20ml of 0.2mol/L Na_2HPO_4 solution and mix them in a 50ml beaker evenly to prepare NaH_2PO_4-Na_2HPO_4 buffer solution A with a buffer ratio of 3∶2. Divide buffer solution A into another 50ml beaker, 25ml per beaker for standby.

Take another two 50ml beakers and put 25ml distilled water into them as blank control solution B for standby.

3. Property determination of buffer solution

Take a small beaker containing 25ml of buffer solution A, use an intelligent acidity meter to measure the initial pH value of the buffer solution and record it. Add 1mol/L HCl to the small beaker for 10 times, and add 0.5ml each time using the pipette. Stir evenly until the reading of acidity meter is stable, measure and record the pH value of buffer solution added with acid each time.

Recalibrate the acidity meter, operate with the same method, measure the pH value of buffer solution after adding 1mol/L NaOH alkali solution for 10 times and record it.

Recalibrate the acidity meter, operate in the same way, use blank control solution B to replace buffer solution A to complete the determination of pH value of the aforementioned acid addition and subtraction solution, and record it.

4. Data analysis and drawing

Enter the pH value data of buffer solution A for 20 times and distilled water for 20 times into excel table respectively Figure 3-1, insert them as data source into "scatter diagram with smooth line and data mark" to obtain curve of pH change, and adjust the horizontal and vertical coordinates and curve name to obtain the figure as shown in Figure 3-2.

Using the property principle of buffer solution to explain the characteristics and differences of the two graphs and complete the experimental report.

Notes

1. When using the intelligent acidity meter to measure the pH value of the solution, after each measurement, the glass bubble at the lower end of the electrode shall be rinsed with distilled water in the washing bottle, and the residual distilled water shall be gently washed off with filter paper scraps to make the measurement results as accurate as possible. Pay attention not to scratch the electrode glass bubble, otherwise the electrode will fail.

2. Using the pipette to measure the added acid-base solution, fix one to measure HCl and the other

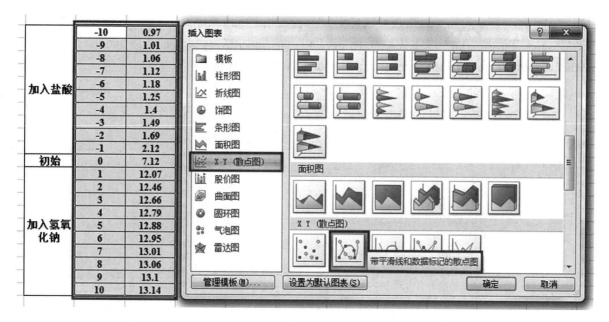

Figure 3-1 Example of inserting "scatter with smooth lines and data markers"

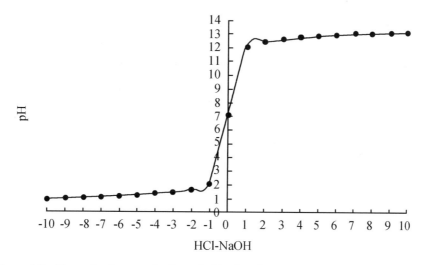

Figure 3-2 Example of curve of adjusted pH value changing with added acid and alkali

to measure NaOH, which can reasonably reduce the repeated cleaning and moistening work caused by changing the pipette, but when using the pipette for the first time, wash it with distilled water for three times, and then moisten it with the acid-base solution to be taken for three times.

3. After the experiment, the intelligent electrode shall be rinsed and put into the protection cap of electrode again to avoid the influence of the complete drying of the electrode glass bubble on the service life. Clean up the test equipment and paper scraps, and keep the test table neat.

Questions

(1) Comparing the curve of buffer solution A and blank control solution B, what is the difference between them? Why are they so different?

(2) In this experiment, NaH_2PO_4–Na_2HPO_4 buffer with a buffer ratio of 3∶2 is used. Is the curve obtained from the experimental data symmetrical along the center? Why? If the buffer ratio is 2∶3, please predict how the curve obtained from the experimental data will change?

实验四 醋酸电离度和电离平衡常数的测定

实验目的

1. **掌握** 容量瓶、滴定管、移液管的基本操作；醋酸溶液的电离度和电离平衡常数的测定方法。

2. **了解** 使用 pH 计的方法。

实验原理

HAc 是一元弱酸，在水溶液中存在下列电离平衡：

$$HAc \rightleftharpoons H^+ + Ac^-$$

$$K_a^\ominus = \frac{[H^+][Ac^-]}{[HAc]} = \frac{c\alpha^2}{1-\alpha}$$

式中 $[H^+]$、$[Ac^-]$、$[HAc]$ 分别是 H^+、Ac^-、HAc 的平衡浓度；c 为醋酸的起始浓度，α 为该浓度醋酸的电离度，$K_a^\ominus$ 为醋酸的电离平衡常数。

实验中采用标准氢氧化钠溶液滴定 HAc 溶液，选择酚酞作指示剂，计算出 HAc 的浓度。测得 HAc 溶液的 pH 后，根据 $pH=-\lg[H^+]$ 计算出 $[H^+]$，再由 $\alpha = \dfrac{[H^+]}{c} \times 100\%$ 计算出电离度 α，再代入 $K_a^\ominus$ 的计算式即可求得电离平衡常数 $K_a^\ominus$。

仪器、试剂及其他

1. **仪器** pHS-3C pH 计、移液管 (5ml, 25ml)、碱式滴定管 (50ml)、容量瓶 (50ml)、烧杯 (50ml)、锥形瓶 (250ml)、滴管等。

2. **试剂** HAc 溶液 (约 0.2mol/L)、标准 NaOH 溶液 (约 0.2mol/L)、标准缓冲溶液、酚酞指示剂。

实验内容

1. **HAc 溶液浓度的标定** 用移液管准确移取 25ml 浓度为 0.2mol/L 左右的 HAc 溶液，置于 250ml 锥形瓶中，加入 2~3 滴酚酞指示剂。用标准 NaOH 溶液滴定至溶液呈现微红色，且 30 秒不褪色，准确记录滴定前后滴定管的读数，计算出消耗的标准 NaOH 溶液的体积，数据记录在表 3-1 中。按照此法滴定 3 次，用 3 次平均值得出 HAc 溶液的准确浓度。

表 3-1　HAc 溶液浓度的标定数据记录表

		1	2	3
c_{NaOH}/(mol/L)				
V_{HAc}/(ml)		25.00	25.00	25.00
V_{NaOH}/(ml)	$V_{始}$/(ml)			
	$V_{终}$/(ml)			
	V_{NaOH}/(ml)			
c_{HAc}/(mol/L)				
$\bar{c}_{HAc}$/(mol/L)				

2. 不同浓度 HAc 溶液的配制　用移液管分别移取 2.5ml、5ml、25ml 的 HAc 溶液置于三个 50ml 容量瓶中，用蒸馏水稀释至刻度，摇匀。连同未稀释的 HAc 溶液可得 $c/20$、$c/10$、$c/2$、c 四种不同浓度的 HAc 溶液，依次编号 1、2、3、4。

3. HAc 溶液 pH 的测定　将 30ml 左右的 1 号、2 号、3 号、4 号 HAc 溶液置于 4 个洁净干燥的 50ml 烧杯中，用 pH 计分别测定它们的 pH，并记录室温。

4. 电离度与电离平衡常数的计算　根据测定的四个浓度的 HAc 溶液的 pH 计算电离度与电离平衡常数，填入表 3-2 中。

表 3-2　HAc 电离度与电离平衡常数的记录表

$t =$ ＿＿＿℃

	c_{HAc}/(mol/L)	pH	$[H^+]$/(mol/L)	α/%	$K_a^{\ominus}$ 测定值	$K_a^{\ominus}$ 平均值
1($c/20$)						
2($c/10$)						
3($c/2$)						
4(c)						

注意事项

1. 预习实验原理、电离度和电离平衡常数的计算方法及影响因素。
2. 预习碱式滴定管、移液管、容量瓶的使用方法，滴定终点的判断方法。
3. 预习 pH 计的使用方法。

思考题

1. 改变醋酸溶液的浓度或温度，则电离度和电离平衡常数有无变化？若有变化，会有怎样的变化？
2. 标定 HAc 溶液时，选用酚酞作指示剂，酚酞的变色范围是多少？应注意什么？可否选用甲基橙作指示剂？

Experiment 4   Determination of Ionization Degree and Equilibrium Constant of Acetic Acid

Objectives

1. Master the basic techniques of volumetric flask, burette and pipette. Master the determination method of ionization degree and equilibrium constant of acetic acid solution.
2. Learn how to use the pH meter.

Principles

HAc is a weak monoacid with the following ionization equilibrium in aqueous solution:

$$HAc \rightleftharpoons H^+ + Ac^-$$

$$K_a^\ominus = \frac{[H^+][Ac^-]}{[HAc]} = \frac{c\alpha^2}{1-\alpha}$$

$[H^+]$, $[Ac^-]$ and $[HAc]$ are the equilibrium concentrations of H^+, Ac^- and HAc, respectively. c is the initial concentration of acetic acid, α is the ionization degree of acetic acid at that concentration, and $K_a^\ominus$ is the ionization equilibrium constant of acetic acid.

In the experiment, the concentration of HAc was calculated by titrating the HAc solution with standard sodium hydroxide solution and using phenolphthalein as indicator. After measuring the pH value of HAc solution, $[H^+]$ can be calculated according to the calculation formula: $pH = -\lg[H^+]$, then the ionization degree can be calculated according to $\alpha = \dfrac{[H^+]}{c} \times 100\%$, and finally the ionization equilibrium constant $K_a^\ominus$ can be obtained.

Equipment, Chemicals, and others

1. Equipment

PHS-3C pH meter, pipette (5ml, 25ml), basic burette (50ml), volumetric flask (50ml), beaker (50ml), conical flask (250ml), dropper.

2. Chemicals

HAc solution (about 0.2mol/L), standard NaOH solution (about 0.2mol/L), standard buffer solution, phenolphthalein indicator.

Experimental Procedure

1. Calibration of HAc solution concentration

25ml 0.2mol/L HAc solution is accurately transferred with a pipette and placed in a 250ml conical flask with 2~3 drops of phenolphthalein indicator added. The standard NaOH solution is dropwise added until the solution turns reddish and retains no fading for 30 seconds. The burette readings before and after titration are recorded accurately, then the volume of the consumed NaOH solution is calculated. The data are recorded in the table 3-1. According to this method, the exact concentration of HAc solution can be obtained by using the mean value of three titrations.

Table 3-1 Data record sheet of calibration of HAc solution concentration

		1	2	3
c_{NaOH}/(mol/L)				
V_{HAc}/(ml)		25.00	25.00	25.00
V_{NaOH}/(ml)	$V_{始}$/(ml)			
	$V_{终}$/(ml)			
	V_{NaOH}/(ml)			
c_{HAc}/(mol/L)				
$\bar{c}_{HAc}$/(mol/L)				

2. Prepare HAc solutions with different concentrations

Transfer 2.5ml, 5ml and 25ml HAc solution into three 50ml volumetric flasks with pipette, dilute with distilled water to the scale, and shake well. Together with the undiluted HAc solution, four HAc solutions of different concentrations, $c/20$, $c/10$, $c/2$ and c, are obtained, numbered 1, 2, 3 and 4 in turn.

3. Determination of pH of HAc solution

30ml HAc solutions No.1, No.2, No.3 and No.4, are placed in four clean and dry 50ml beakers, and their pH are measured by pH meter respectively, and room temperature is recorded.

4. Calculation of ionization degree and ionization equilibrium constant

The ionization degree and ionization equilibrium constant can be calculated according to the pH of HAc solution of different concentrations. write them in table 3-2.

Table 3-2 Record sheet of the ionization degree and ionization equilibrium constant of HAc solution

$t =$ ___ °C

	c_{HAc}/(mol/L)	pH	[H$^+$]/(mol/L)	α/%	$K_a^\ominus$	
					measurement	mean
1($c/20$)						
2($c/10$)						
3($c/2$)						
4(c)						

Notes

1. Preview experimental principles, calculation methods and influencing factors of ionization degree

and ionization equilibrium constants.

2. Preview the use method of basic burette, pipette, volumetric flask, and determination the end-point of titration.

3. Preview the usage of pH meter.

 Questions

1. If the concentration or temperature of acetic acid solution is changed, is there any change in ionization degree and ionization equilibrium constant? If there is a change, what will it be?

2. Phenolphthalein was used as an indicator in the calibration the HAc solution, what is the color changing range of phenolphthalein? What precautions should be taken? Can methyl orange be used as indicator?

实验五 氯化铅溶度积常数的测定

实验目的

1. **掌握** 离子交换法测定难溶电解质溶度积常数的原理和方法。
2. **学习** 离子交换树脂的使用方法。
3. 进一步训练酸碱滴定的基本操作。

实验原理

离子交换树脂是一种具有可供离子交换的活性基团的高分子化合物。具有酸性交换基团 (如磺酸基—SO_3H、羧基—$COOH$),能和阳离子进行交换的叫阳离子交换树脂。具有碱性交换基团 (如—NH_3Cl),能和阴离子进行交换的叫阴离子交换树脂。本实验采用钠 (Na^+) 型强酸性阳离子交换树脂 (活性基团为—SO_3Na),使用时需要 1mol/L HCl 溶液浸泡或淋洗使之转型,即用 H^+ 把 Na^+ 交换下来,制得氢 (H^+) 型树脂 (活性基团为—SO_3H)。

例如:一定量的饱和 $PbCl_2$ 溶液加入离子交换柱中与 H^+ 型阳离子树脂充分接触,交换反应如下:

$$2R—SO_3H + PbCl_2 =\!=\!= (R—SO_3)_2Pb + 2HCl$$

交换出的 HCl 可用已知浓度的 NaOH 标准溶液来滴定。根据反应方程式即可算出 $PbCl_2$ 饱和溶液的浓度,从而求得 $PbCl_2$ 的溶解度和溶度积。计算公式如下:

$$c_{NaOH} \times V_{NaOH} = c_{HCl} \times V_{HCl} = 2c_{PbCl_2} \times V_{PbCl_2}$$

$$K_{sp}^{\ominus}(PbCl_2) = [Pb^{2+}][Cl^-]^2 = c_{PbCl_2} \times (2c_{PbCl_2})^2 = 4c_{PbCl_2}^3 = 4\left(\frac{c_{NaOH} \times V_{NaOH}}{2V_{PbCl_2}}\right)^3$$

已用过的离子交换树脂可用 0.1mol/L HNO_3 溶液进行淋洗或浸泡,使树脂重新转化为酸型,这个过程称之为再生。树脂在使用前需用蒸馏水浸泡 24~48 小时,使用后也需要浸泡在水中以保持活性。

仪器、试剂及其他

1. **仪器** 碱式滴定管 (25ml),移液管 (10ml),烧杯 (100ml),锥形瓶 (250ml),量筒 (100ml、10ml),漏斗,温度计。
2. **试剂** $PbCl_2$(s,分析纯),溴百里酚蓝 (0.1%),NaOH 标准溶液 (约 0.05mol/L),HNO_3 (0.1mol/L),阳离子交换树脂。
3. **其他** 滤纸,玻璃纤维或泡沫塑料,广泛 pH 试纸。

实验内容

1. **$PbCl_2$ 饱和溶液的制备** 将 0.5g $PbCl_2$ 固体溶于 50ml 蒸馏水 (经煮沸除去 CO_2,并冷却至室温),充分搅拌,使溶液达到平衡,静置,记录室温,过滤,所用漏斗和盛接容器 (100ml 烧杯)都必须是干燥的。

2. 装柱 用碱式滴定管作为离子交换柱,在底部放入少量玻璃纤维或泡沫塑料,以防止树脂漏出。用小烧杯量取约 10ml 阳离子交换树脂(已经过转型或再生,并用清水调成"糊状"),用蒸馏水洗至中性(用 pH 试纸检验),注入交换柱内。如果水太多,可打开螺旋夹,让水流出,直到液面略高于离子交换树脂,夹紧螺旋夹。以上操作中,一定要使树脂始终浸在溶液中,勿使溶液流干,否则气泡进入树脂床中,将影响离子交换的进行。若出现气泡,可加入少量蒸馏水,使液面高出树脂,反复上下倒转滴定管,赶走气泡。

3. 交换和洗涤 用移液管精密量取 10.00ml $PbCl_2$ 饱和溶液,注入离子交换柱中。控制交换柱流出液的速度,每分钟 20~25 滴,用洁净锥形瓶盛接流出液。待 $PbCl_2$ 饱和溶液液面接近树脂层上表面时,用蒸馏水少量多次注入交换柱,洗涤交换树脂,直至流出液呈中性为止(用 pH 试纸检验),流出液仍用同一只锥形瓶盛接。在整个交换和洗涤过程中,应注意勿使流出液损失。

4. 滴定 在流出液中加入 2~3 滴溴百里酚蓝指示剂,用 NaOH 标准溶液滴定到终点(溶液由黄色转为蓝色,pH=6.2~7.6)。精确记录下 NaOH 溶液的用量。

5. 再生 用 5ml 0.1mol/L HNO_3 浸泡树脂使之再生。

6. 计算 运用公式 $K_{sp}^{\ominus}(PbCl_2) = 4\left(\dfrac{c_{NaOH} \times V_{NaOH}}{2V_{PbCl_2}}\right)^3$ 计算氯化铅溶度积常数。

注意事项

1. 在洗涤、交换的过程中,树脂交换柱中的液面始终高于树脂面,在整个交换和洗涤过程中,应注意及时加入蒸馏水,防止树脂暴露在空气中。
2. 制备 $PbCl_2$ 饱和溶液时,要用煮沸除去 CO_2 的热水溶解 $PbCl_2$ 固体。
3. $PbCl_2$ 饱和溶液通过交换柱后,要用蒸馏水洗至中性,并且不允许流出液有所损失。

思考题

1. 离子交换过程中,为什么要控制液体的流速?为什么要始终保持树脂交换柱中的液面高于树脂面?
2. 为何再生时用 HNO_3 替代 HCl?

Experiment 5　Determination of Solubility Product Constant of PbCl$_2$

Objectives

1. Grasp the principle and method of measuring the solubility product constant of slightly soluble electrolyte by ion exchange method.
2. Learn how to use ion exchange resins.
3. Practice the basic operation of acid-base titration.

Principles

Ion exchange resin is a kind of high molecular compound with active groups available for ion exchange. The resins that have acid exchange groups (such as the sulfonate—SO$_3$H and carboxyl—COOH), and can be exchanged by cations are called cation exchange resins. The resins with basic exchange groups (such as—NH$_3$Cl) that can be exchanged by anions are called the anion exchange resins. In this experiment, a strong acid cation exchange resin [sodium (Na$^+$) type strong acidic cation exchange resin] is used. Its active group is—SO$_3$Na. It needs to be transformed by being soaked or rinsed with 1mol/L HCl solution, that is, replaced with H$^+$ to produce hydrogen (H$^+$) type resin (the active group is —SO$_3$H).

For example, a certain amount of saturated PbCl$_2$ solution is added to the ion exchange column to make sufficient contact with the H$^+$ type cationic resin. The exchange reaction is as following:

$$2R—SO_3H + PbCl_2 = (R—SO_3)_2Pb + 2HCl$$

The exchanged HCl can be titrated with a standard NaOH solution of known concentration. According to the reaction equation, the concentration of PbCl$_2$ saturated solution can be calculated, and the solubility and solubility product constant of PbCl$_2$ can be obtained. The calculation formulas are as following:

$$c_{NaOH} \times V_{NaOH} = c_{HCl} \times V_{HCl} = 2c_{PbCl_2} \times V_{PbCl_2}$$

$$K_{sp}^{\ominus}(PbCl_2) = [Pb^{2+}][Cl^-]^2 = c_{PbCl_2} \times (2c_{PbCl_2})^2 = 4c_{PbCl_2}^3 = 4\left(\frac{c_{NaOH} \times V_{NaOH}}{2V_{PbCl_2}}\right)^3$$

The used ion exchange resin can be rinsed or soaked with 0.1mol/L HNO$_3$ solution, so that the resin can be converted into an acid form again. This process is called regeneration. The resin needs to be soaked with distilled water for 24~48 hours before use, and after experiment it needs to be soaked in water to maintain activity.

Equipment, Chemicals and Others

1. Equipment

Basic burettes (25ml), pipettes (10ml), beakers (100ml), erlenmeyer flasks (250ml), measuring cylinders (100ml, 10ml), funnel, thermometers.

2. Chemicals

PbCl$_2$ (s, analytical grade), bromothymol blue (0.1%), standard NaOH solution (about 0.05mol/L), HNO$_3$ (0.1mol/L), cation exchange resin.

3. Others

Filter paper, glass fiber or foam plastic, universal pH test paper.

Experimental Procedures

1. Preparation of PbCl$_2$ Saturated Solution

0.5g PbCl$_2$ is dissolved into 50ml distilled water (remove CO$_2$ by boiling, and cool to room temperature). Stir thoroughly to make the solution reach equilibrium, and then let the solution stand quietly. Record the room temperature. Filter the PbCl$_2$ saturated solution and note that the funnel and receiving container (100ml beaker) must be dry!

2. Packing

Use a basic burette as an ion exchange column and put a small amount of glass fiber or foam plastic at the bottom to prevent resin leakage. The transformed or regenerated cation exchange resin was used to make a "paste" with water. And then we take out 10ml with a beaker and rinse it to neutrality with distilled water (check with pH paper). Then the resin is injected into the column. If there is too much water, we should open the screw clamp and let the water flow out until the liquid level is slightly higher than the ion exchange resin. In the above operation, the resin must always be immersed in the solution. The solution should not be drained to avoid air bubbles entering the resin bed, which will affect the ion exchange. If air bubbles appear, a small amount of distilled water should be added to make the liquid level higher than the resin. And then we should turn the burette upside down repeatedly to drive the air bubbles out.

3. Swap and Wash

10.00ml PbCl$_2$ saturated solution is taken out by pipette into the ion exchange column. The flow rate of the exchange column is controlled at the speed of 20~25 drops per minute. Use a clean erlenmeyer flask to receive the effluent. When the surface of the PbCl$_2$ saturated solution reaches the upper surface of the resin layer, a small amount of distilled water is injected into the exchange column to rinse the exchange resin until the effluent is neutral (check with pH paper). Do not lose the effluent during the entire exchange and washing process.

4. Titration

Add 2~3 drops of bromothymol blue indicator to the effluent and titrate to the end point with NaOH solution (the solution turns from yellow to blue, pH=6.2~7.6). Record the volume of the NaOH solution accurately.

5. Regeneration

Regenerate the resin by soaking it with 5ml 0.1mol/L HNO$_3$.

6. Calculation

Calculate the solubility product constant of PbCl$_2$ with the following formula:

$$K_{sp}^{\ominus}(PbCl_2) = 4\left(\frac{c_{NaOH} \times V_{NaOH}}{2V_{PbCl_2}}\right)^3$$

Notes

1. During the entire washing and exchange process, the liquid level in the resin exchange column is always higher than the resin surface. And distilled water should be added in time to prevent the resin from being exposed to the air.

2. When preparing the $PbCl_2$ saturated solution, the $PbCl_2$ solids are dissolved with hot water with CO_2 removed by boiling.

3. After the $PbCl_2$ saturated solution pass through the exchange column, wash the column with distilled water until it is neutral, and the effluent should not be lost.

Questions

(1) Why should the flow rate of the liquid be controlled during the ion exchange process? Why does the liquid level in the resin exchange column always keep higher than the resin?

(2) Why is HNO_3 used instead of HCl during the regeneration?

实验六　氧化还原反应与电极电势

实验目的

1. **掌握**　电极电势的概念并观察电极电势对氧化还原反应的影响。
2. **了解**　原电池的装置。
3. **观察**　浓度、酸度对电极电势的影响；浓度、酸度、温度、催化剂对氧化还原反应的方向、产物、速度的影响。

实验原理

1. 氧化还原反应的实质是物质在反应过程中发生了电子的转移或偏移。氧化剂在反应中得电子，还原剂失电子。氧化剂、还原剂的相对强弱，可用它们的氧化型物质及对应的还原型物质所组成的氧化还原电对的电极电势的相对大小来衡量。电极电势值越高，则氧化还原电对中氧化态物质的氧化能力越强，那么该氧化态物质就是一种相对较强的氧化剂；电极电势值越低，则氧化还原电对中还原态物质的还原能力越强，那么该还原态物质就是一种相对较强的还原剂。

2. 利用氧化还原反应而产生电流的装置称为原电池。原电池的电动势等于两个电极电势之差，即：$E_{池}=E_{正}-E_{负}$。只有 $E_{池}>0$，氧化还原反应才能正向进行。因此根据电极电势的大小，还可以判断氧化还原反应进行的方向。

3. 浓度、酸度、温度均影响电极电势的数值。它们之间的关系可用 Nernst 方程式表示：

$$E = E^{\ominus} + \frac{RT}{nF}\ln\frac{c_{Ox}^a}{c_{Red}^b} \quad 或 \quad E = E^{\ominus} + \frac{0.0592}{n}\lg\frac{c_{Ox}^a}{c_{Red}^b}$$

仪器、试剂及其他

1. **仪器**　试管，烧杯，伏特计，恒温水浴锅。

2. **试剂**

酸：1mol/L H_2SO_4，2mol/L HCl，浓 HCl，3mol/L H_2SO_4，6mol/L HAc，0.1mol/L $H_2C_2O_4$。

碱：浓 $NH_3 \cdot H_2O$，6mol/L NaOH。

盐：0.1mol/L KI，0.1mol/L $FeCl_3$，0.1mol/L KBr，0.5mol/L $ZnSO_4$，0.5mol/L $CuSO_4$，0.5mol/L $FeSO_4$，0.4mol/L $K_2Cr_2O_7$，0.01mol/L $KMnO_4$，0.1mol/L Na_2SO_3，0.1mol/L $MnSO_4$，0.1mol/L $AgNO_3$，固体 MnO_2，固体 $(NH_4)_2S_2O_8$。

3. **其他**　电极（锌片、铜片、铁片、碳棒），导线，鳄鱼夹，砂纸，淀粉-碘化钾试纸。

实验内容

1. **电极电势与氧化还原反应的关系**　在试管中加入 0.1ml 0.1mol/L KI 溶液和 2 滴 0.1mol/L $FeCl_3$ 溶液，混匀后加入 0.5ml CCl_4，充分振荡，观察 CCl_4 层的颜色变化并解释之。

用0.1mol/L KBr溶液代替0.1mol/L KI溶液，进行同样的实验。观察实验现象并解释之。

根据以上实验的结果，定性比较 Br_2/Br^-、I_2/I^- 和 Fe^{3+}/Fe^{2+} 三个电对的电极电势的相对大小，并指出最强的氧化剂和最强的还原剂，并说明电极电势与氧化还原反应进行方向有何关系。写出有关反应方程式。

2. 浓度和酸度对电极电势的影响

（1）浓度的影响　取两个50ml的烧杯，在一个烧杯中加入30ml 0.5mol/L $ZnSO_4$ 溶液，在溶液中插入 Zn 片，另一个烧杯中加入30ml 0.5mol/L $CuSO_4$ 溶液，在溶液中插入 Cu 片。将两种溶液用盐桥连接起来，用导线将 Zn 片和 Cu 片分别与伏特计的负极和正极相连，测量两电极间电压（图3-3）。

取出盐桥，向 $ZnSO_4$ 溶液中边滴加浓氨水边搅拌，至生成的沉淀完全溶解后，重新放入盐桥，观察电压有何变化并解释之。写出有关反应方程式。

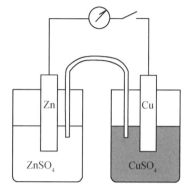

图3-3　原电池装置示意图

取出盐桥，向 $CuSO_4$ 溶液中边滴加浓氨水边搅拌，至生成的沉淀完全溶解形成深蓝色溶液后，重新放入盐桥，观察电压有何变化并解释之。写出有关反应方程式。

（2）酸度的影响　取两个50ml的烧杯，在一个烧杯中加入30ml 0.5mol/L $FeSO_4$ 溶液，在溶液中插入 Fe 片，在另一个烧杯中加入30ml 0.4mol/L $K_2Cr_2O_7$ 溶液，在溶液中插入碳棒。将两种溶液用盐桥连接起来，用导线将 Fe 片和碳棒分别与伏特计的负极和正极相连，测量两电极间电压。

向 $K_2Cr_2O_7$ 溶液中慢慢加入 1mol/L H_2SO_4 溶液，观察电压有何变化。再向 $K_2Cr_2O_7$ 溶液中逐滴加入 6mol/L NaOH 溶液，观察电压又有何变化。

3. 酸度对氧化还原产物的影响　取三支试管，各加入2滴 0.01mol/L $KMnO_4$ 溶液，然后在第一支试管中加入 0.5ml 1mol/L H_2SO_4 溶液，第二支试管中加入 0.5ml 蒸馏水，第三支试管中加入 0.5ml 6mol/L NaOH 溶液，再向三支试管中各加入 0.5ml 0.1mol/L Na_2SO_3 溶液。观察3支试管中颜色的变化，写出反应方程式并解释之。

4. 浓度对氧化还原反应方向的影响　向试管中加入少许固体 MnO_2 和10滴 2mol/L HCl 溶液，将湿润的淀粉-碘化钾试纸放在试管口。观察有无 Cl_2 生成。

用浓 HCl 代替 2mol/L HCl 溶液进行同样的实验。比较两次的实验结果并解释之。写出有关反应方程式。

5. 酸度、温度和催化剂对氧化还原反应速度的影响

（1）酸度的影响　向2支试管中各加入5滴 0.1mol/L KBr 溶液，然后在1支试管中加入10滴 3mol/L H_2SO_4 溶液，另1支试管中加入10滴 6mol/L HAc 溶液，再各加入1滴 0.01mol/L $KMnO_4$ 溶液。观察并比较2支试管中紫色褪去的快慢。写出反应方程式并解释之。

（2）温度的影响　向2支试管中各加入5滴 0.1mol/L $H_2C_2O_4$ 溶液和1滴 0.01mol/L $KMnO_4$ 溶液，摇匀。将其中1支试管水浴加热数分钟，另1支不加热。观察2支试管中紫色褪去的快慢。写出反应方程式并解释之。

（3）催化剂的影响　向2支试管中各加入10滴 3mol/L H_2SO_4 溶液、1滴 0.1mol/L $MnSO_4$ 溶液和少量 $(NH_4)_2S_2O_8$ 固体，振荡使其溶解。然后往1支试管中加入1~2滴 0.1mol/L $AgNO_3$ 溶液，另1支不加。微热，观察2支试管中颜色的变化，写出反应方程式并解释之。

注意事项

1. 电极、导线接头及鳄鱼夹等都必须用砂纸打干净,以免接触不良。
2. $FeSO_4$ 溶液和 Na_2SO_3 溶液易被空气中氧气氧化,因而需临用新制。
3. $(NH_4)_2S_2O_8$ 为强氧化剂,实验结束后的废液应回收到指定容器。

思考题

1. 根据标准电极电势如何判断氧化剂和还原剂的相对强弱?如何判断氧化还原反应进行的方向?
2. 实验室用 MnO_2 和盐酸制备 Cl_2 时,为什么用浓盐酸而不用稀盐酸?
3. 介质对 $KMnO_4$ 氧化性有何影响?
4. 通过实验,你熟悉了哪些氧化剂和还原剂?它们的产物是什么?

Experiment 6 Redox Reaction and Electrode Potential

Objectives

1. Master the concept of electrode potential and observe the effect of electrode potential on the redox reaction.
2. Understand the device of galvanic cell.
3. Observe the effect of concentration and acidity on electrode potential. Observe the effects of concentration, acidity, temperature, and catalyst on the direction, product, and speed of the redox reaction.

Principles

1. The essence of redox reaction is transfer or migration of electrons. In the redox reaction, the oxidant gains electrons and reductant loses electrons. The relative strength of the oxidants and reductants can be measured by electrode potential of redox couples formed by their oxidizing substances and the corresponding reducing substances. The higher the value of electrode potential is, the stronger the oxidative ability of oxidizing substance in redox couple is, so this oxidizing substance is a stronger oxidant. The lower the value of electrode potential is, the stronger the reduction ability of reducing substance in redox couple is, so this reducing substance is a stronger reductant.

2. The device which can generate electricity by redox reaction is called galvanic cell. The electromotive force of galvanic cell is equal to the positive electrode potential minus the negative electrode potential, that is, $E_{cell}=E_{positive}-E_{negative}$. Only when $E_{cell}>0$, the redox reaction can proceed in the forward direction. Therefore, the direction of redox reaction can be determined according to the electrode potential.

3. The value of electrode potential is affected by concentration, acidity, and temperature. The relationship between them can be expressed by Nernst equation.

$$E=E^{\ominus}+\frac{RT}{nF}\ln\frac{c_{Ox}^a}{c_{Red}^b} \quad 或 \quad E=E^{\ominus}+\frac{0.0592}{n}\lg\frac{c_{Ox}^a}{c_{Red}^b}$$

Equipment, Chemicals and Others

1. Equipment

Test tube, beaker, voltmeter, thermostatic water bath.

2. Chemicals

Acid: 1mol/L H_2SO_4, 2mol/L HCl, concentrated hydrochloric acid, 3mol/L H_2SO_4, 6mol/L HAc, 0.1mol/L $H_2C_2O_4$.

Base: strong ammonia, 6mol/L NaOH.

Salt: 0.1mol/L KI, 0.1mol/L $FeCl_3$, 0.1mol/L KBr, 0.5mol/L $ZnSO_4$, 0.5mol/L $CuSO_4$, 0.5mol/L $FeSO_4$, 0.4mol/L $K_2Cr_2O_7$, 0.01mol/L $KMnO_4$, 0.1mol/L Na_2SO_3, 0.1mol/L $MnSO_4$, 0.1mol/L $AgNO_3$, MnO_2(s), $(NH_4)_2S_2O_8$(s).

3. Others

Electrode (zinc slices, copper slices, iron slices, carbon rod), wire, crocodile clip, sandpaper, potassium iodide-starch test paper.

Experimental Procedures

1. Relationship between Electrode Potential and Redox Reaction

(1) The solution of KI (0.1mol/L, 0.1mL) and $FeCl_3$ (0.1mol/L, two drops) were added to the test tube and well-mixed. Then, the liquid of CCl_4 (0.5ml) was added to the mixture. What's the color change in CCl_4 layer after fully shaking the test tube? Explain the experimental phenomenon.

(2) The same experiment was performed with the solution of KBr (0.1mol/L) instead the solution of KI (0.1mol/L). Observe and explain the experimental phenomenon.

What's the order of the relative values of electrode potential of the three redox couples which include Br_2/Br^-, I_2/I^-, and Fe^{3+}/Fe^{2+} after qualitative comparison based on the results of above experiments? What's the the strongest oxidant and reductant? What's the relationship between electrode potential and the direction of redox reaction? Write down the equations for the reaction.

2. Effects of Concentration and Acidity on Electrode Potential

2.1 Concentration

(1) The solution of $ZnSO_4$ (0.5mol/L, 30mL) was added to a 50ml beaker and zinc slice was inserted into solution. The solution of $CuSO_4$ (0.5mol/L, 30ml) was added to another 50ml beaker and copper slice was inserted into solution. The two solutions were connected by a salt bridge. Zinc slice was connected to the negative electrode of voltmeter with wire. Copper slice was connected to the positive electrode of voltmeter with wire. The voltage between two electrodes was measured. The device of galvanic cell was shown in the Figure 3-3.

(2) Concentrated ammonia was added dropwise to the solution of $ZnSO_4$ after taking out the salt bridge until the formed precipitate was completely dissolved. How does the voltage change after putting in salt bridge? Explain the experimental phenomenon and write down the equations for the reaction.

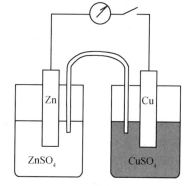

Figure 3-3 The diagram of galvanic cell device

(3) Concentrated ammonia was added dropwise to the solution of $CuSO_4$ after taking out the salt bridge until the formed precipitate was completely dissolved to form dark blue solution. How does the voltage change after putting in salt bridge? Explain the experimental phenomenon and write down the equations for the reaction.

2.2 Acidity

(1) The solution of $FeSO_4$ (0.5mol/L, 30ml) was added to a 50ml beaker and iron slice was inserted into solution. The solution of $K_2Cr_2O_7$ (0.4mol/L, 30ml) was added to another 50ml beaker and carbon rod was inserted into solution. The two solutions were connected by a salt bridge. Iron slice was connected to the negative electrode of voltmeter with wire. Carbon rod was connected to the positive electrode of

voltmeter with wire. The voltage between two electrodes was measured.

(2) The solution of H$_2$SO$_4$ (1mol/L) was slowly added to the solution of K$_2$Cr$_2$O$_7$ and observe the voltage change. Then, the solution of NaOH (6mol/L) was added dropwise to the solution of K$_2$Cr$_2$O$_7$ and observe the voltage change.

3. Effect of Acidity on the Redox Reaction Products

The solution of KMnO$_4$ (0.01mol/L, two drops) was added to three test tube. The solution of H$_2$SO$_4$ (1mol/L, 0.5ml) was added to the first test tube. The distilled water (0.5ml) was added to the second test tube. The solution of NaOH (6mol/L, 0.5ml) was added to the third test tube. The solution of Na$_2$SO$_3$ (0.1mol/L, 0.5ml) was added to every test tube. How does the color change in three test tubes? Explain the experimental phenomenon and write down the equations for the reaction.

4. Effect of Concentration on the Direction of Redox Reaction

(1) A little soild MnO$_2$ and the solution of HCl (2mol/L, ten drops) were added to the test tube. Wet potassium iodide-starch test paper was placed in the test tube mouth. Does chlorine produce?

(2) The same experiment was performed with concentrated hydrochloric acid instead the solution of HCl (2mol/L). Compare and explain the results of two experiments. Write down the equations for the reaction.

5. Effects of Acidity, Temperature and Catalyst on the Rate of Redox Reaction

5.1 Acidity

The solution of KBr (0.1mol/L, five drops) was added to each of two test tubes. The solution of H$_2$SO$_4$ (3mol/L, ten drops) was added to the first test tube. The solution of HAc (6mol/L, ten drops) was added to the second test tube. Then, the solution of KMnO$_4$ (0.01mol/L, one drop) was added to every test tube. Which test tube faded quickly in purple? Explain the experimental phenomenon and write down the equations for the reaction.

5.2 Temperature

The solution of H$_2$C$_2$O$_4$ (0.1mol/L, five drops) and the solution of KMnO$_4$ (0.01mol/L, one drop) were added to each of two test tubes and shaken up. One test tube was heated for several minutes in water bath and the other was not heated. Which test tube faded quickly in purple? Explain the experimental phenomenon and write down the equations for the reaction.

5.3 Catalyst

The solution of H$_2$SO$_4$ (3mol/L, ten drops), the solution of MnSO$_4$ (0.1mol/L, one drop) and a little solid (NH$_4$)$_2$S$_2$O$_8$ were added to each of two test tubes and shaken up into liquid. Then, the solution of AgNO$_3$ (0.1mol/L, 1~2 drops) was added to one test tube and the other was not added. How does the color change in two test tube after heating slightly? Explain the experimental phenomenon and write down the equations for the reaction.

Notes

1. Electrodes, wire connection and crocodile clip must be cleaned with sandpaper to prevent bad contact.

2. The solutions of FeSO$_4$ and Na$_2$SO$_3$ need to be newly prepared because they are easily oxidized by oxygen in the air.

3. Since the substance of (NH$_4$)$_2$S$_2$O$_8$ is a strong oxidant, the waste liquid in experiment should be recycled to the designated container.

Questions

1. How to judge the relative strength of the oxidant and reductant according to the standard electrode potential? How to judge the direction of redox reaction?

2. Why concentrated hydrochloric acid was used instead of dilute hydrochloric acid when preparing chlorine with MnO_2 and hydrochloric acid in laboratory?

3. How does the medium affect the oxidative properties of $KMnO_4$?

4. What oxidants and reductants did you know through experiments? What are their corresponding products?

（姚　军）

实验七　配合物的生成、性质与应用

实验目的

1. **熟悉**　加热、过滤和试管反应的基本操作。
2. **了解**　几种不同类型的配合物的生成。配合物与简单化合物和复盐的区别。影响配位平衡的因素。螯合物的形成条件。

实验原理

1. 由中心原子与配体按一定的组成和空间构型以配位键结合所形成的化合物称为配位化合物。配合物中配离子在水中的行为与弱电解质相似，只能极少量地解离出中心离子和配体。而在简单化合物或复盐的溶液中，各种离子都能表现出游离离子的性质。由此可以区分配合物与简单化合物和复盐。

2. 配离子在水溶液中存在着配位平衡，其平衡常数用 K_s 表示，称为配合物的稳定常数。例如：

$$Cu^{2+}(aq) + 4NH_3(aq) \rightleftharpoons [Cu(NH_3)_4]^{2+}(aq)$$

依据化学平衡原理，其平衡常数表达式为：

$$K_s = \frac{[Cu(NH_3)_4^{2+}]}{[Cu^{2+}][NH_3]^4}$$

3. 配位平衡与其他化学平衡一样，受外界条件的影响，如加入沉淀剂、氧化剂、还原剂或改变介质的酸度，平衡都将发生移动。

4. 中心原子与多齿配体形成的环状配合物称为螯合物。很多金属螯合物具有特征颜色。有些特征反应常用来作为金属离子的鉴定反应。

仪器、试剂及其他

1. **仪器**　试管，漏斗，玻璃棒，离心试管，白瓷点滴板，离心机。

2. **试剂**

酸：3mol/L H_2SO_4，1mol/L HCl。

碱：2mol/L $NH_3 \cdot H_2O$，6mol/L $NH_3 \cdot H_2O$，0.1mol/L NaOH，1mol/L NaOH。

盐：0.1mol/L $CuSO_4$，0.1mol/L $HgCl_2$，0.1mol/L KI，0.1mol/L $BaCl_2$，0.1mol/L $AgNO_3$，0.1mol/L NaCl，0.1mol/L KBr，0.1mol/L $Na_2S_2O_3$，0.1mol/L $FeCl_3$，0.5mol/L $FeCl_3$，1mol/L 枸橼酸钠，0.1mol/L KSCN，0.1mol/L EDTA，0.1mol/L $FeSO_4$，4mol/L NH_4F。

3. **其他**　95% 乙醇，CCl_4，0.25% 邻菲罗啉，滤纸。

实验内容

1. 配合物的生成 在试管中加入 2ml 0.1mol/L $CuSO_4$ 溶液,逐滴加入 2mol/L $NH_3 \cdot H_2O$,至产生的沉淀变为深蓝色溶液为止。然后加入 4ml 乙醇,震荡均匀,放置 2~3 分钟,过滤析出的结晶 $[Cu(NH_3)_4SO_4 \cdot H_2O]$ 用少量的乙醇洗 1~2 次,保留以供后面实验备用。记录产品的性质并写出有关反应方程式。

在试管中加入 3 滴 0.1mol/L $HgCl_2$ 溶液,逐滴加入 0.1mol/L KI 溶液,至产生的沉淀完全溶解为止。写出有关反应方程式。

2. 配合物与简单化合物或复盐的区别 在两支试管中各加入少量 $[Cu(NH_3)_4SO_4 \cdot H_2O]$,滴入几滴水溶解。在一支试管中加入 2 滴 0.1mol/L NaOH 溶液,在另一支试管中加入 2 滴 0.1mol/L $BaCl_2$ 溶液。观察实验现象并解释,写出有关反应方程式。

在两支试管中各加入 5 滴 0.1mol/L $CuSO_4$ 溶液。在一支试管中加入 2 滴 0.1mol/L NaOH 溶液,在另一支试管中加入 2 滴 0.1mol/L $BaCl_2$ 溶液。观察实验现象并解释,写出有关反应方程式。

用实验说明硫酸铁铵是复盐,铁氰化钾是配合物,写出实验步骤并用实验验证。

3. 配位平衡的移动

(1) 配位平衡与沉淀溶解平衡 在离心试管中加入 5 滴 0.1mol/L $AgNO_3$ 溶液和 5 滴 0.1mol/L NaCl 溶液,离心后弃去清液,然后加入 6mol/L $NH_3 \cdot H_2O$ 至沉淀刚好溶解为止。

往上述溶液中加 1 滴 0.1mol/L NaCl 溶液,观察是否有白色沉淀生成,再加 1 滴 0.1mol/L KBr 溶液,观察沉淀的颜色。继续加入 0.1mol/L KBr 溶液,至不再产生沉淀为止。离心后弃去清液,在沉淀中加入 0.1mol/L $Na_2S_2O_3$ 溶液直至沉淀刚好溶解为止。

往上述溶液中加入 1 滴 0.1mol/L KBr 溶液,观察有无 AgBr 沉淀生成,再加 1 滴 0.1mol/L KI 溶液,观察有无 AgI 沉淀生成。

根据以上实验结果,讨论沉淀溶解和配位平衡的关系,并比较 AgCl、AgBr、AgI 稳定常数的大小以及 $[Ag(NH_3)_2]^+$、$[Ag(S_2O_3)_2]^{3-}$ 配离子的稳定性大小。解释实验现象并写出各步反应式。

(2) 配位平衡与氧化还原反应 在两支试管各加入 0.5ml 0.5mol/L $FeCl_3$ 溶液,在一支试管中逐滴加入 4mol/L NH_4F 至溶液呈无色。然后在两支试管中分别加入 5 滴 0.1mol/L KI 溶液和 0.5ml CCl_4,振荡均匀。观察两支试管中 CCl_4 层的颜色。解释实验现象并写出反应方程式。

(3) 配位平衡与介质酸碱性 在试管中加入 5 滴 0.1mol/L $CuSO_4$ 溶液,再逐滴加入 6mol/L $NH_3 \cdot H_2O$ 直到沉淀完全溶解。然后逐滴加入 3mol/L H_2SO_4,观察溶液颜色变化,是否有沉淀生成。继续加入 3mol/L H_2SO_4 至溶液显酸性,观察溶液颜色变化。解释上述现象并写出反应方程式。

在试管中加入 1ml 0.1mol/L $FeCl_3$ 溶液,再继续加入 1ml 1mol/L 枸橼酸钠溶液。观察颜色变化。然后将溶液分到两支试管中,向一支试管中滴加 1mol/L NaOH 溶液使成碱性,在另一支试管中滴加 1mol/L HCl 溶液使成酸性,观察两支试管颜色有何不同。

4. 螯合物的形成 取 10 滴 0.1mol/L $FeCl_3$ 溶液于试管中,再滴加一滴 0.1mol/L KSCN 溶液。另取少量 $[Cu(NH_3)_4SO_4 \cdot H_2O]$ 于另一试管中,滴入几滴水溶解。然后向两支试管中分别滴加 0.1mol/L EDTA 溶液,观察实验现象并解释之。

在白瓷点滴板上滴一滴 0.1mol/L $FeSO_4$ 溶液,再加 3 滴 0.25% 邻菲罗啉溶液,观察实验现象并解释之。

注意事项

1. $HgCl_2$ 毒性很大,使用时要注意安全。实验结束后的废液应回收到指定容器。

2. 制备 [Cu(NH$_3$)$_4$]SO$_4$ 时，首先要将 CuSO$_4$ 固体全部溶解后才能加 NH$_3$·H$_2$O，而且必须加浓 NH$_3$·H$_2$O。

思考题

1. 影响配位平衡的因素有哪些？
2. 配合物与复盐的主要区别是什么？
3. 实验中所用的 EDTA 是什么物质？它与单基配体有何不同？

Experiment 7 Formation, Properties and Application of Complexes

Objectives

1. Familiar with the basic operation of heating, filtering and test tube reaction.
2. Understand the formation of several different types of complexes. Compare the difference between complexes and simple compounds or double salts. To understand the factors that affect coordination equilibrium. To understand the formation conditions of chelates.

Principles

1. Complexes are formed by a central atom and ligands in a certain composition and spatial configuration with coordination bonds. Like weak electrolytes in water, complexes can only dissociate the central atom and ligands in a very small amount. Various ions in the solution of simple compounds or double salts can show the properties of free ions. This property can be used to distinguish complex from simple compounds and double salt.

2. There exist coordination equilibriums in aqueous solution of complex ions. The constant of coordination equilibrium is represented by K_s, which is called the stability constant of complex. For example:

$$Cu^{2+}(aq) + 4NH_3(aq) \rightleftharpoons [Cu(NH_3)_4]^{2+}(aq)$$

According to the principle of chemical equilibrium, the expression of coordination equilibrium constant is as follows:

$$K_s = \frac{[Cu(NH_3)_4^{2+}]}{[Cu^{2+}][NH_3]^4}$$

3. Like other chemical equilibrium, coordination equilibrium is affected by external conditions. When adding precipitants, oxidants, reducing agents, or changing the acidity of the medium, the coordination equilibrium will be moved.

4. Cyclic complex formed by the central atom and multidentate ligands is called chelate. Many metal chelates have characteristic colors. Some characteristic reactions are commonly used as identification reactions for metal ions.

Equipment, Chemicals and Others

1. Equipment

Test tube, funnel, glass rod, centrifuge tube, white porcelain spot plate, centrifuge.

2. Chemicals

Acid: 3mol/L H_2SO_4, 1mol/L HCl.

Base: 2mol/L $NH_3 \cdot H_2O$, 6mol/L $NH_3 \cdot H_2O$, 0.1mol/L NaOH, 1mol/L NaOH.

Salt: 0.1mol/L $CuSO_4$, 0.1mol/L $HgCl_2$, 0.1mol/L KI, 0.1mol/L $BaCl_2$, 0.1mol/L $AgNO_3$, 0.1mol/L NaCl, 0.1mol/L KBr, 0.1mol/L $Na_2S_2O_3$, 0.1mol/L $FeCl_3$, 0.5mol/L $FeCl_3$, 1mol/L sodium citrate, 0.1mol/L KSCN, 0.1mol/L EDTA, 0.1mol/L $FeSO_4$, 4mol/L NH_4F.

3. Others

Ethanol(95%), CCl_4, 1,10-Phenanthroline hydrate (0.25%), Filter paper.

Experimental Procedures

1. Formation of Complexes

(1) The solution of $CuSO_4$ (0.1mol/L, 2ml) was added to the test tube. Then the solution of $NH_3 \cdot H_2O$ (2mol/L) was added dropwise into solution until the precipitate became dark blue solution. Crystal of $[Cu(NH_3)_4SO_4 \cdot H_2O]$ was precipitated after adding ethanol (4ml) to solution, shaking uniformly and stand for 2~3 mintures. Crystal precipitation was filtered and washed 1~2 times with small amount of ethanol. Record the properties of the product and write down the reaction equation.

(2) The solution of $HgCl_2$ (0.1mol/L, three drops) was added to the test tube. Then the solution of KI (0.1mol/L) was added dropwise into solution until the precipitate was completely dissolved. Write down the reaction equation.

2. Difference between Complexes and Simple Compounds or Double Salts

(1) A small amount of $[Cu(NH_3)_4SO_4 \cdot H_2O]$ was added to each of two test tubes and dissolved in a few drops of water. The solution of NaOH (0.1mol/L, two drops) was added to one test tube. Then the solution of $BaCl_2$ (0.1mol/L, two drops) was added to another test tube. Observe and explain the experimental phenomenon. Write down the reaction equations.

(2) The solution of $CuSO_4$ (0.1mol/L, five drops) was added to each of two test tubes. Then, the solution of NaOH (0.1mol/L, two drops) was added to one test tube. The solution of $BaCl_2$ (0.1mol/L, two drops) was added to another test tube. Observe and explain the experimental phenomenon. Write down the reaction equations.

(3) The experiment is designed to show that the compound of $(NH_4)Fe(SO_4)_2$ and $K_3[Fe(CN)_6]$ is a double salt and complex, respectively. Write down the experimental procedure and verify them by experiment.

3. Shift of Coordination Equilibrium

3.1 Coordination Equilibrium and Precipitation Dissolution

The solution of $AgNO_3$ (0.1mol/L, five drops) and NaCl (0.1mol/L, five drops) were added to centrifuge tube. The supernatant was discarded after centrifugation. Then the solution of $NH_3 \cdot H_2O$ (6mol/L) was added to the centrifuge tube until the precipitate was completely dissolved.

Is there any white precipitate generated when adding the solution of NaCl (0.1mol/L, one drop) in the above solution? What color is the precipitate when adding the solution of KBr (0.1mol/L, one drop)? Then the solution of KBr (0.1mol/L) was continually added until no precipitation occurred. The supernatant was discarded after centrifugation. Then the solution of $Na_2S_2O_3$ (0.1mol/L) was added to the centrifuge tube until the precipitate was completely dissolved.

Is there any precipitate of AgBr formed when adding the solution of KBr (0.1mol/L, one drop) in the above solution? Then observe whether the precipitate of AgI is formed after adding the solution of KI (0.1mol/L, one drop).

What is the relationship between precipitation dissolution and coordination equilibrium according to the above experimental results? Compare the stability constant value of AgCl、AgBr、AgI. Judge the complex ion stability of $[Ag(NH_3)_2]^+$ and $[Ag(S_2O_3)_2]^{3-}$. Explain the experimental phenomenon and write down the reaction equations.

3.2 Coordination Equilibrium and Redox Reaction

The solution of $FeCl_3$ (0.5mol/L, 0.5ml) was added to each of two test tubes. Then the solution of NH_4F (4mol/L) was added dropwise into one test tube until the color of solution became colorless. Add the solution of KI (0.1mol/L, five drops) and CCl_4 to each test tube respectively and shake them evenly. Observe the color of CCl_4 layer in two test tubes. Explain the experimental phenomenon and write down the reaction equations.

3.3 Coordination Equilibrium and the Acidity-basicity of Meduim

(1) The solution of $CuSO_4$ (0.1mol/L, five drops) was added to the test tube. Then the solution of $NH_3·H_2O$ (6mol/L) was added dropwise into solution until the precipitate was completely dissolved. Observe the color change of solution and whether the precipitate was formed after adding the solution of H_2SO_4 (3mol/L). The solution of H_2SO_4 (3mol/L) was continually added until the solution was acidic. Observe the color change of solution. Explain the above experimental phenomenon and write down the reaction equations.

(2) The solution of $FeCl_3$ (0.1mol/L, 1ml) was added to the test tube. Then the solution of sodium citrate (1mol/L, 1ml) was added into solution. The above solution was divided into two test tubes after observing the color change. The solution of NaOH (1mol/L) was added dropwise into one test tube until the solution was alkaline. The solution of HCl (1mol/L) was added dropwise into another test tube until the solution was acidic. Observe the color difference between the two test tubes.

4. Formation of Chelates

(1) The solution of KSCN (0.1mol/L, one drops) was added to the test tube which contained the solution of $FeCl_3$ (0.1mol/L, ten drops). A small amount of $[Cu(NH_3)_4SO_4·H_2O]$ was added to the test tube and dissolved in a few drops of water. Then the solution of EDTA (0.1mol/L) was added dropwise into each of the two test tubes. Observe and explain the experimental phenomenon.

(2) The solution of $FeSO_4$ (0.1mol/L, one drop) was added to white porcelain spot plate. Then add the solution of 1,10-Phenanthroline hydrate (0.25%, three drops). Observe and explain the experimental phenomenon.

Notes

1. Since the substance of $HgCl_2$ is very toxic, the waste liquid in experiment should be recycled to the designated container.

2. When preparing $[Cu(NH_3)_4]SO_4$, the solid of $CuSO_4$ is completely dissolved before adding strong ammonia.

Questions

1. What are the factors that affect coordination equilibrium?
2. What's the main differences between complex and double salt?
3. What's EDTA in experiment? What's the difference between EDTA and monodentate ligand?

实验八 银氨配离子配位数的测定

实验目的

1. 测定 $[Ag(NH_3)_n]^+$ 配离子的配位数 n，并计算稳定常数 $K_S^\ominus$。
2. 巩固深化配位平衡及溶度积原理。
3. 学习如何处理实验数据和作图。

实验原理

在硝酸银溶液中加入过量的氨水，即生成稳定的银氨配离子 $[Ag(NH_3)_n]^+$，再往溶液中加入溴化钾溶液，直至刚刚开始有 AgBr 沉淀（浑浊）出现为止。这时混合溶液中同时存在着配位平衡和沉淀平衡。

配位平衡：$Ag^+ + nNH_3 \rightleftharpoons [Ag(NH_3)_n]^+$

$$K_S^\ominus = \frac{[Ag(NH_3)_n]^+}{[Ag^+][NH_3]^n} \tag{3-1}$$

沉淀平衡：$AgBr(s) \rightleftharpoons Ag^+ + Br^-$

$$K_{sp}^\ominus(AgBr) = [Ag^+][Br^-] \tag{3-2}$$

式 (3-1)×(3-2)，得：

$AgBr(s) + nNH_3 \rightleftharpoons [Ag(NH_3)_n]^+ + Br^-$

$$K^\ominus = \frac{[Ag(NH_3)_n^+][Br^-]}{[NH_3]^n} = K_S^\ominus \times K_{sp}^\ominus(AgBr) \tag{3-3}$$

整理 (3-3) 式，得

$$[Ag(NH_3)_n^+][Br^-] = K^\ominus[NH_3]^n \tag{3-4}$$

将式 (3-4) 两边取对数，得直线方程：

$$\lg\{[Ag(NH_3)_n^+][Br^-]\} = n\lg[NH_3] + \lg K^\ominus \tag{3-5}$$

以 $\lg\{[Ag(NH_3)_n^+][Br^-]\}$ 为纵坐标，$\lg[NH_3]$ 为横坐标作图，所得直线的斜率即为 $[Ag(NH_3)_n]^+$ 的配位数 n。

根据直线在 y 轴的截距可求得 $K^\ominus$。$K_{sp}^\ominus(AgBr)$ 已知，可代入 (3-3) 式求得 $[Ag(NH_3)_n]^+$ 的稳定常数 $K_S^\ominus$。

$[Br^-]$、$[NH_3]$ 和 $[Ag(NH_3)_n^+]$ 皆指平衡时的浓度，它们可近似计算如下：

设每份混合溶液中，最初取用的 $AgNO_3$ 溶液体积为 V_{Ag^+}（各份相同），初浓度为 $[Ag^+]_0$，每份加入的 $NH_3 \cdot H_2O$ 溶液（大量过量）和 KBr 溶液的体积分别为 V_{NH_3} 和 V_{Br^-}，其初浓度分别为 $[NH_3]_0$ 和 $[Br^-]_0$。混合溶液的总体积为 $V_总$，$V_总 = V_{Br^-} + V_{NH_3} + V_{Ag^+}$，则混合后并达平衡时：

$$[Br^-] = [Br^-]_0 \times \frac{V_{Br^-}}{V_总} \tag{3-6}$$

$$[NH_3] = [NH_3]_0 \times \frac{V_{NH_3}}{V_总} \tag{3-7}$$

$$[Ag(NH_3)_n^+] = [Ag^+]_0 \times \frac{V_{Ag^+}}{V_总} \tag{3-8}$$

仪器、试剂及其他

1. **仪器** 移液管（20ml）；锥形瓶（250ml）；滴定管（2×50ml）。
2. **试剂** 0.010mol/L AgNO$_3$，0.010mol/L KBr，2.00mol/L 氨水（应新鲜配制）。

实验内容

1. 用移液管吸取 25.00ml 0.010mol/L AgNO$_3$ 溶液放入 1 个 250ml 锥形瓶中。
2. 用碱式滴定管量取 20.00ml 2.00mol/L 氨水溶液加入上述锥形瓶中。
3. 从酸式滴定管中向锥形瓶逐滴加 0.010mol/L KBr 溶液，直至开始产生的 AgBr 浑浊不再消失时，停止滴定，记下所加入的 KBr 溶液的体积 V_{Br^-} 和溶液的总体积 $V_总$。
4. **重复操作** 向同一锥形瓶中继续加入 5.00ml 2.00mol/L 氨水，则两次所加氨水的体积之和为 25.00ml，然后继续逐滴加 KBr 溶液，同样至刚出现不消失的浑浊为止。记下两次累计用去 KBr 溶液的体积 V_{2,Br^-} 和溶液总体积 $V_总$。

用同上方法，按照表 3-3 中氨水的累计用量继续进行 4 次滴定实验。记录加入 KBr 溶液的累计体积 V_{3,Br^-}、V_{4,Br^-}、V_{5,Br^-}、V_{6,Br^-}，填入表 3-3 中。

5. 数据记录

表 3-3　银氨配离子配位数的测定数据记录表

滴定序号	1	2	3	4	5	6
V_{Ag^+} (ml)	25.00	25.00	25.00	25.00	25.00	25.00
V_{NH_3} (ml)	20.00	25.00	30.00	35.00	40.00	45.00
V_{Br^-} (ml)						
$V_总$ (ml)						
[Br$^-$]（mol/L）						
[NH$_3$] (mol/L)						
[Ag(NH$_3$)$_n^+$] (mol/L)						
lg{[Ag(NH$_3$)$_n^+$][Br$^-$]}						
lg[NH$_3$]						

6. 结果处理

（1）以 lg{[Ag(NH$_3$)$_n^+$][Br$^-$]} 为纵坐标，lg[NH$_3$] 为横坐标作图。
（2）从图求得配位数 n，并从公式（3-3）求出 [Ag(NH$_3$)$_n$]$^+$ 的 $K_S^\ominus$（文献值：$K_{sp,AgBr}^\ominus = 4.1 \times 10^{-13}$）。

注意事项

1. 配制混合溶液时，最后加氨水（防止氨水挥发掉）。氨水必须是新标定的，若为放置过久的，要重新标定。
2. 反应一定要达平衡（振摇后沉淀不消失）后再观察终点，且每次浑浊度要一致。
3. 实验完毕将取 AgNO$_3$ 溶液用的移液管、锥形瓶用实验剩余的氨水洗涤以防 Ag$^+$ 转化成 Ag。

思考题

1. 在计算平衡浓度 $[Br^-]$、$[NH_3]$ 和 $[Ag(NH_3)_n^+]$ 时，为什么可以忽略以下情况？

（1）生成 AgBr 沉淀时消耗掉的 Br^- 和 Ag^+。

（2）配离子 $[Ag(NH_3)_n]^+$ 离解出的 Ag^+。

（3）生成配离子 $[Ag(NH_3)_n]^+$ 时消耗的 NH_3。

2. 滴定时，若 KBr 溶液加过量了，有无必要弃去锥形瓶中的溶液重新开始？

3. 本实验用的锥形瓶为什么必须是干燥的？为什么滴定过程中始终不能加水？

Experiment 8 Determining Coordination Number of $[Ag(NH_3)_n]^+$

Objectives

1. Determine the coordination number (n) and the stability constant ($K_s^\ominus$) of $[Ag(NH_3)_n]^+$.
2. Deepen the cognition of coordination equilibrium principle and solubility product principle.
3. Learn how to deal with the data and how to draw the graph.

Principle

The stable complex ion consisting of Ag^+ and NH_3 can be formed when excess ammonia water is added into silver nitrate solution. Nevertheless, silver bromide (AgBr) is precipitated when a potassium bromide solution is added into the solution of the complex. The coordination equilibrium and the precipitation equilibrium coexist in this system.

The coordination Equilibrium: $Ag^+ + nNH_3 \rightleftharpoons [Ag(NH_3)_n]^+$

$$K_s^\ominus = \frac{[Ag(NH_3)_n]^+}{[Ag^+][NH_3]^n} \tag{3-1}$$

The precipitation Equilibrium: $AgBr(s) \rightleftharpoons Ag^+ + Br^-$

$$K_{sp}^\ominus(AgBr) = [Ag^+][Br^-] \tag{3-2}$$

(3-1)×(3-2):

$AgBr(s) + nNH_3 \rightleftharpoons [Ag(NH_3)_n]^+ + Br^-$

$$K^\ominus = \frac{[Ag(NH_3)_n^+][Br^-]}{[NH_3]^n} = K_s^\ominus \times K_{sp}^\ominus(AgBr) \tag{3-3}$$

Then:

$$[Ag(NH_3)_n^+][Br^-] = K^\ominus[NH_3]^n \tag{3-4}$$

Make the logarithms at both sides of the equation(3-4), we can obtain a liner equation：

$$\lg\{[Ag(NH_3)_n^+][Br^-]\} = n\lg[NH_3] + \lg K^\ominus \tag{3-5}$$

Make the graph (assign y-coordinate to $\lg\{[Ag(NH_3)_n^+][Br^-]\}$, x-coordinate to $\lg[NH_3]$, and intercept to $\lg K^\ominus$) and then we can get the value of the slope(n) which is also the value of the coordination number of $[Ag(NH_3)_n]^+$ (take the closest integer).

Put intercept ($\lg K^\ominus$) into Equation (3-3), as we have known $K_{sp}^\ominus(AgBr)$, then we have the stability constant $K_s^\ominus$ of $[Ag(NH_3)_n]^+$ complex ion.

$[Br^-]$, $[NH_3]$ and $[Ag(NH_3)_n^+]$ represent the equilibrium concentrations and can be approximately calculated by the following method:

In each portion of the mixture, if the volume of silver nitrate solution we initially took is the same as V_{Ag^+} and the concentration is represented as $[Ag^+]_0$. If the volume of ammonia water (substantial excess) we added sequentially into each portion is V_{NH_3} and the concentration is $[NH_3]_0$. If the volume of potassium bromide solution we titrate into each portion is V_{Br^-} and the concentration is $[Br^-]_0$. And if the total volume of mixture is V_t, then, when equilibrium is reached in the mixture, we have:

$$[Br^-]=[Br^-]_0 \times \frac{V_{Br^-}}{V_t} \tag{3-6}$$

$$[NH_3] = [NH_3]_0 \times \frac{V_{NH_3}}{V_t} \tag{3-7}$$

$$[Ag(NH_3)_n^+] = [Ag^+]_0 \times \frac{V_{Ag^+}}{V_t} \tag{3-8}$$

Equipment, Chemicals and Others

1. Equipment

Transfer pipette (20ml); Erlenmeyer flask (250ml); burettes (2×50ml).

2. Chemicals

0.010mol/L Silver nitrate, 0.010mol/L potassium bromide, 2.00mol/L ammonia solution.

Experimental Procedures

1. At first, 25.00ml silver nitrate solution with the concentration 0.010mol/L is pipetted into a 250ml Erlenmeyer flask through the transfer pipette.

2. Secondly, 20.00ml ammonia water with the concentration 2.00mol/L is added into the same Erlenmeyer flask through the basic burette.

3. Thirdly, the complex ion solution in the Erlenmeyer flask is titrated by a 0.010mol/L potassium bromide solution through the acid burette until the precipitate of silver bromide appears. Then stop the titration and write down the volume V_{Br^-} of potassium bromide solution consumed and the ultimate volume V_t of the solution in that flask.

4. Repeat operation: 5.00ml ammonia water with the concentration 2.00mol/L is added into the same Erlenmeyer flask through the basic burette continually, so the sum of the volumes of ammonia water added twice is 25.00ml, then the complex ion solution in the Erlenmeyer flask is titrated by a 0.010mol/L potassium bromide solution through the acid burette until the precipitate of silver bromide appears. Then stop the titration and write down the accumulated volume V_{2,Br^-} of potassium bromide solution consumed twice and the ultimate accumulated volume V_t of the solution in that flask.

In the same way, 4 more times operations are repeated continually according to the accumulated volume of ammonia solutions in the table 3-3. Write down the accumulated volume V_{Br^-} (such as V_{3,Br^-}, V_{4,Br^-}, V_{5,Br^-} and V_{6,Br^-}) into table 3-3.

5. Records

Table 3-3 Data record sheet of determining Coordination Number of $[Ag(NH_3)_n]^+$

Number	1	2	3	4	5	6
V_{Ag^+} (ml)	25.00	25.00	25.00	25.00	25.00	25.00
V_{NH_3} (ml)	20.00	25.00	30.00	35.00	40.00	45.00
V_{Br^-} (ml)						
V_t (ml)						

Continued

Number	1	2	3	4	5	6
$[Br^-]$ (mol/L)						
$[NH_3]$ (mol/L)						
$[Ag(NH_3)_n^+]$ (mol/L)						
$\lg\{[Ag(NH_3)_n^+][Br^-]\}$					.	
$\lg[NH_3]$						

6. Data Process

6.1 The graph with y-coordinate assigned to $\lg\{[Ag(NH_3)_n^+][Br^-]\}$ and x-coordinate assigned to $\lg[NH_3]$ should be made.

6.2 The value of *n* can be attained from the graph and the value of $K_s^\ominus$ will be calculated out from Equation (8-3). The reference value: $K_{sp,AgBr}^\ominus = 4.1 \times 10^{-13}$.

Notes

1. During the concoction for the mixture, ammonia solution is added finally for it is volatile. Ammonia water should be a newly calibrated solution and must be re-calibrated if it is laid aside for a long time.

2. It is preferable that the turbidity degrees at the endpoints are the same in each titration with respect of the parallel principle.

3. At last, pipettes and Erlenmeyer flasks contaminated with silver nitrate should be washed with the left ammonia solution because silver ion is converted to metal silver easily.

Questions

(1) Why could the following situations be ignored when the equilibrium concentrations of $[Br^-]$, $[NH_3]$ and $[Ag(NH_3)_n^+]$ are calculated?

a. The amount of Br^- and Ag^+ used to produce the precipitate of silver bromide.

b. The amount of Ag^+ dissociated from $[Ag(NH_3)_n]^+$.

c. The amount of ammonia used to produce the complex ion of $[Ag(NH_3)_n]^+$.

(2) If potassium bromide solution was added in excess during titration, is it the best disposal that the solution in the Erlenmeyer flask was thrown away for a turn-over-a-new-leaf?

(3) Why does the Erlenmeyer flask used in this experiment have to be dry? Why can't add water during the titration process?

（林　舒）

实验九 铬、锰、铁

实验目的

1. **熟悉** 铬、锰、铁各种主要氧化态之间的转化。
2. **了解** 铬（Ⅲ）、锰（Ⅱ）、铁（Ⅱ、Ⅲ）氢氧化物的生成和性质。铬（Ⅵ）、锰（Ⅶ）化合物的氧化还原性以及介质对氧化还原反应的影响。

实验原理

铬（Cr）、锰（Mn）、铁（Fe）依次属于ⅥB、ⅦB和Ⅷ族元素，在化合物中Cr、Mn的最高价态和族数相等。Fe的最高价则小于族数。Cr常见氧化态为 +3、+6；Mn 为 +2、+4、+6、+7；Fe 为 +2、+3。

$Cr(OH)_3$ 灰绿色，两性；$Mn(OH)_2$ 白色，碱性；$Fe(OH)_2$ 白色，碱性；$Fe(OH)_3$ 棕色，两性极弱；$Mn(OH)_2$ 和 $Fe(OH)_2$ 极易被空气氧化为 $MnO(OH)_2$（棕黑）和 $Fe(OH)_3$（棕）。

由 Cr（Ⅲ）氧化成 Cr（Ⅵ），需加入氧化剂，且在碱性介质中进行，如：

$$2CrO_2^- + 3H_2O_2 + 2OH^- = 2CrO_4^{2-} + 4H_2O$$

而 Cr（Ⅵ）还原成 Cr（Ⅲ），需加入还原剂，且在酸性介质中进行，如

$$Cr_2O_7^{2-} + 3S^{2-} + 14H^+ = 2Cr^{3+} + 3S + 7H_2O$$

铬酸盐和重铬酸盐在溶液中存在下列平衡：

$$2CrO_4^{2-} + 2H^+ = Cr_2O_7^{2-} + H_2O$$

加酸或碱可使平衡移动。一般多酸盐溶解度比单酸盐大，故在 $K_2Cr_2O_7$ 溶液中加入 Pb^{2+}，实际生成 $PbCrO_4$ 黄色沉淀。

Mn（Ⅵ）由 MnO_2 和强碱在氧化剂 $KClO_3$ 的作用下加强热而制得，绿色锰酸钾溶液极易歧化：

$$3K_2MnO_4 + 4HAc = 2KMnO_4 + MnO_2 + 4KAc + 2H_2O$$

K_2MnO_4 可被 Cl_2 氧化成 $KMnO_4$。

$KMnO_4$ 是强氧化剂，它的还原产物随介质酸碱性不同而异。在酸性溶液中 MnO_4^- 被还原成 Mn^{2+}，在中性溶液中被还原为 MnO_2，在强碱性介质中被还原成 MnO_4^{2-}。

Fe^{3+} 和 Fe^{2+} 均易和 CN^- 形成配合物，Fe^{3+} 与 $[Fe(CN)_6]^{4-}$ 反应、Fe^{2+} 与 $[Fe(CN)_6]^{3-}$ 反应均生成蓝色沉淀或溶胶，前者称普鲁士蓝，后者称滕氏蓝。最近，被证明它们的结构相同，为 $[KFe^{Ⅱ}(CN)_6Fe^{Ⅲ}]$。

仪器、试剂及其他

1. **仪器** 试管，试管架，洗瓶。
2. **试剂**

酸：（2mol/L，6mol/L）H_2SO_4，2mol/L HAc，3% H_2O_2。

碱：（0.5、2、6mol/L）NaOH，KOH（s）。

盐：0.1mol/L $KCr(SO_4)_2$，0.1mol/L $K_2Cr_2O_7$，0.1mol/L $Pb(NO_3)_2$，0.1mol/L KSCN，2mol/L

$(NH_4)_2S$，0.1mol/L $MnSO_4$，0.01mol/L $KMnO_4$，0.1mol/L Na_2SO_3，0.1mol/L $FeCl_3$，0.1mol/L KI，CCl_4，0.1mol/L $K_4[Fe(CN)_6]$，0.1mol/L $K_3[Fe(CN)_6]$，0.1mol/L $(NH_4)_2Fe(SO_4)_2$，$KClO_3$（s），MnO_2（s），$(NH_4)_2Fe(SO_4)_2·6H_2O$（s），PbO_2（s）。

3. **其他** $CHCl_3$ 或淀粉溶液，氯水。

实验内容

1. **铬（Ⅲ）化合物**

（1）$Cr(OH)_3$ 的产生 取两支试管，分别注入 0.1mol/L $KCr(SO_4)_2$ 5滴和 0.5mol/L NaOH 1~2滴，观察灰绿色 $Cr(OH)_3$ 沉淀的生成。

（2）$Cr(OH)_3$ 的两性 向上述两支试管中分别滴加 2mol/L H_2SO_4 和 2mol/L NaOH，有何变化？

（3）Cr（Ⅲ）被氧化 向上面制得的 $NaCrO_2$ 溶液中加入 3% H_2O_2 3~4滴并加热，观察现象的变化，写出反应式。

2. **铬（Ⅵ）化合物**

（1）溶液中 CrO_4^{2-} 与 $Cr_2O_7^{2-}$ 间的平衡移动 取4滴 0.1mol/L $K_2Cr_2O_7$，观察其颜色，加入数滴 2mol/L NaOH，观察其颜色变化；再加入数滴 2mol/L H_2SO_4，颜色又有何变化？

取4滴 0.1mol/L $K_2Cr_2O_7$ 溶液，滴加 2滴 0.1mol/L $Pb(NO_3)_2$，观察 $PbCrO_4$ 沉淀的生成。

（2）Cr(Ⅵ)的氧化性 取4滴 0.1mol/L $K_2Cr_2O_7$ 溶液，加 2滴 2mol/L H_2SO_4 酸化，再加 2滴 2mol/L $(NH_4)_2S$ 溶液，微热，观察现象及颜色变化。

3. **锰（Ⅱ）化合物**

（1）$Mn(OH)_2$ 的生成和性质 在 10滴 0.1mol/L $MnSO_4$ 溶液中，加 5滴 2mol/L NaOH，立即观察现象（不振摇），放置后再观察现象有何变化？

（2）Mn（Ⅱ）的被氧化 往试管中加入少许 PbO_2（s）、1ml 6mol/L H_2SO_4 及 1滴 0.1mol/L $MnSO_4$，将试管用小火加热，小心振荡，静置后观察溶液转为紫红色，写出反应式，并用电极电势说明之。

4. **锰（Ⅵ）化合物**

（1）K_2MnO_4 的生成 在一干燥小试管中放入一小粒 KOH 和约等体积的 $KClO_3$ 晶体（尽可能少取），加热至熔结在一起后，再加入少许 MnO_2，加热熔融，至熔结后，使试管口稍低于管底部，强热至熔块呈绿色，放置，待冷后加 2ml 水振荡使溶，溶液应呈绿色。写出反应式。

（2）K_2MnO_4 的歧化 取少量上面自制的 K_2MnO_4 溶液，加入稀醋酸，观察溶液颜色的变化和沉淀的生成。写出离子反应式。

5. **锰（Ⅶ）化合物** 取三支试管各加入 1滴 0.1mol/L $KMnO_4$ 溶液，其中第一支加入 5滴 1mol/L H_2SO_4，第二支加入 5滴蒸馏水，第三支加入 5滴 6mol/L NaOH，然后分别加 1滴 0.1mol/L Na_2SO_3 溶液，观察各试管所发生的现象。写出反应式，并做出介质对 $KMnO_4$ 还原产物影响的结论。

6. **铁（Ⅱ）化合物** 向试管中加入 2ml 蒸馏水，用 1~2滴 2mol/L H_2SO_4 酸化，然后向其中加入几粒硫酸亚铁铵晶体。在另一支试管中煮沸 1ml 2mol/L NaOH，迅速加到硫酸亚铁铵的溶液中去（不要振摇），观察现象。然后振摇，静置片刻，观察沉淀颜色的变化，解释每步操作的原因和现象的变化。写出有关离子反应式。

7. **铁（Ⅲ）化合物**

（1）在 4滴 0.1mol/L $FeCl_3$ 溶液中加 3滴 2mol/L NaOH，观察现象并写出反应式。

（2）在 4滴 0.1mol/L $FeCl_3$ 溶液中加 2滴 0.1mol/L KI 溶液，观察现象，设法检验所得产物是什么？写出离子反应式。

8. Fe(Ⅱ)、Fe(Ⅲ)的配合物

（1）在 5 滴 0.1mol/L FeCl$_3$ 溶液中，加 1 滴 0.1mol/L K$_4$[Fe(CN)$_6$]，观察普鲁士蓝蓝色沉淀（或溶胶）的形成。

（2）在 5 滴 0.1mol/L (NH$_4$)$_2$Fe(SO$_4$)$_2$ 溶液中，加 1 滴 0.1mol/L K$_3$[Fe(CN)$_6$]，观察滕氏蓝蓝色沉淀（或溶胶）的形成。

（3）在 5 滴 0.1mol/L (NH$_4$)$_2$Fe(SO$_4$)$_2$ 溶液中，加入 1 滴 2mol/L H$_2$SO$_4$ 及 0.1mol/L KSCN 溶液 1 滴，观察有无现象变化？然后再滴加 3%H$_2$O$_2$ 溶液 1 滴，观察颜色的变化。写出离子反应式。

注意事项

1. 试验 Cr^{3+} 还原性时，H$_2$O$_2$ 为氧化剂，有时溶液会出现褐红色，这是由于生成过铬酸钠的缘故。

2. 在酸性溶液中，MnO$_4^-$ 被还原成 Mn^{2+}，有时会出现 MnO$_2$ 的棕色沉淀，这是因溶液的浓度不够以及 KMnO$_4$ 过量，与生成的 Mn^{2+} 反应所致：

$$2MnO_4^- + 3Mn^{2+} + 2H_2O = 5MnO_2 \downarrow + 4H^+$$

思考题

1. 如何鉴定 Cr^{3+} 或 Mn^{2+} 的存在？
2. 怎样存放 KMnO$_4$ 溶液？为什么？
3. 试用两种方法实现 Fe^{2+} 和 Fe^{3+} 的相互转化。

Experiment 9 Chromium, Manganese and Iron

Objectives

1. Familiar with the transformation among the main oxidation states of chromium, manganese and iron.

2. Understand the preparation and properties of chromium (Ⅲ), manganese (Ⅱ) and iron (Ⅱ, Ⅲ) hydroxides. To understand oxidation-reduction properties of compounds chromium (Ⅵ), manganese (Ⅶ). Know how the medium affects these oxidation-reduction reactions.

Principles

Chromium (Cr), manganese (Mn) and iron (Fe) are elements of Group Ⅵ B, Ⅶ B and Ⅷ, respectively. In the compounds, the max valence state of Cr and Mn and group numbers are equal. The max valence state of Fe is less than its group number. Oxidations states of Cr are +3 and +6. Those of Mn are +2, +4, +6, and +7. Those of Fe are +2 and +3.

$Cr(OH)_3$ is green and amphoteric; $Mn(OH)_2$ is white and basic; $Fe(OH)_2$ is white and basic; $Fe(OH)_3$ is brown and weak amphoteric; $Mn(OH)_2$ and $Fe(OH)_2$ are easily oxidized to $MnO(OH)_2$ (brownish-black) and $Fe(OH)_3$ (brown) by the air.

In alkaline medium, Cr(Ⅲ) can be oxidized to Cr(Ⅵ) by oxidizing agent, as follows:

$$2CrO_2^- + 3H_2O_2 + 2OH^- = 2CrO_4^{2-} + 4H_2O$$

In acid medium, Cr(Ⅵ) can be reduced to Cr(Ⅲ) by reducing agent, as follows:

$$Cr_2O_7^{2-} + 3S^{2-} + 14H^+ = 2Cr^{3+} + 3S + 7H_2O$$

There is an equilibrium reaction for chromate and dichromate.

$$2CrO_4^{2-} + 2H^+ = Cr_2O_7^{2-} + H_2O$$

The equilibrium can shift when adding acid or alkaline. Generally, solubility of polyacid salt is higher than that of monoacid salt, so yellow sediment $PbCrO_4$ will be formed when Pb^{2+} is added to $K_2Cr_2O_7$ solution.

When MnO_2 is fused with powerful alkaline in the presence of oxidizing agent $KClO_3$, Mn(Ⅵ) is obtained, the green potassium manganate solution is easy to be disproportionated:

$$3K_2MnO_4 + 4HAc = 2KMnO_4 + MnO_2 + 4KAc + 2H_2O$$

K_2MnO_4 can be oxidized to $KMnO_4$ by Cl_2.

$KMnO_4$ is powerful oxidants and the reduction products vary with the acid-base properties of the media. In acid solutions, MnO_4^- can be reduced to Mn^{2+}; in neutral solutions to MnO_2 and in alkaline solutions to MnO_4^{2-}.

Both Fe^{3+} and Fe^{2+} can form complexes with CN^-. The reaction products of Fe^{3+} and $[Fe(CN)_6]^{4-}$, Fe^{2+} and $[Fe(CN)_6]^{3-}$ are both blue sediment or sol, the former is prussian blue while the latter is Tengs blue. Recently, it is proved that their structures are the same, namely, $[KFe^{Ⅱ}(CN)_6Fe^{Ⅲ}]$.

Equipment, Chemicals and Others

1. Equipment

Test tube, test tube rack, washing bottle

2. Chemicals

Acids: (2mol/L, 6mol/L) H_2SO_4, 2mol/L HAc, 3% H_2O_2.

Alkalis: (0.5, 2, 6mol/L) NaOH, KOH (s).

Salt: 0.1mol/L $KCr(SO_4)_2$, 0.1mol/L $K_2Cr_2O_7$, 0.1mol/L $Pb(NO_3)_2$, 0.1mol/L KSCN, 2mol/L $(NH_4)_2S$, 0.1mol/L $MnSO_4$, 0.01mol/L $KMnO_4$, 0.1mol/L Na_2SO_3, 0.1mol/L $FeCl_3$, 0.1mol/L KI, CCl_4, 0.1mol/L $K_4[Fe(CN)_6]$, 0.1mol/L $K_3[Fe(CN)_6]$, 0.1mol/L $(NH_4)_2Fe(SO_4)_2$, $KClO_3(s)$, $MnO_2(s)$, $(NH_4)_2Fe(SO_4)_2 \cdot 6H_2O(s)$, $PbO_2(s)$.

Else: $CHCl_3$ or starch solution, chlorine water.

Experimental Procedures

1. Cr (III) compounds

1.1 Preparation of $Cr(OH)_3$

Add 5 drops of 0.1mol/L $KCr(SO_4)_2$ and 1~2 drops of 0.5mol/L NaOH to two test tubes respectively. Observe the production of gray green sediments $Cr(OH)_3$.

1.2 Amphotericity of $Cr(OH)_3$

Add 2mol/L H_2SO_4 and 2mol/L NaOH solutions to two test tubes referred above respectively. Observe the change.

1.3 Oxidation of Cr (III)

Add 3~4 drops of 3% H_2O_2 to $NaCrO_2$ solution prepared above and heat. Observe the change and write out the equation of the reaction.

2. Cr (VI) compounds

2.1 Equilibrium shift between CrO_4^{2-} and $Cr_2O_7^{2-}$ in solution

Take one test tube, add 4 drops of 0.1mol/L $K_2Cr_2O_7$ solution and observe the color, add some drops of 2mol/L NaOH and observe the color again, then add some drops of 2mol/L H_2SO_4, what is the change of the color?

Take one test tube, add 4 drops of 0.1mol/L $K_2Cr_2O_7$ solution and 2 drops of 0.1mol/L $Pb(NO_3)_2$, observe the formation of sediment $PbCrO_4$.

2.2 Oxidizability of Cr (VI)

Add 2 drops of 2mol/L H_2SO_4 to 4 drops of 0.1mol/L $K_2Cr_2O_7$ solution, then add 2 drops of 2mol/L $(NH_4)_2S$ solution. Observe the phenomenon and the change of color when heated slightly.

3. Mn (II) compounds

3.1 Preparation and properties of $Mn(OH)_2$

Add 5 drops of 2mol/L NaOH to 10 drops of 0.1mol/L $MnSO_4$ solution, observe the phenomenon immediately (without shaken), place and observe the change of phenomenon.

3.2 Oxidation of Mn (II)

Add a little $PbO_2(s)$, 1ml 6mol/L H_2SO_4 and a drop of 0.1mol/L $MnSO_4$ in test tube, heat the tube slightly and shake carefully, place for a moment and observe that the solution change to prunosus. Then write down the equation and explain it with electrode potential.

4. Mn (Ⅵ) compounds

4.1 Formation of K_2MnO_4

A pill of KOH and a same bulk of $KClO_3$ crystal (as little as possible) are added to a small dry test tube. Heat till they are fused together, add a little MnO_2, continue heating. After they are all fused together, make the upper end of tube lower than the bottom. Heat fusion powerfully till it turns green, and place and cool it down. Then add 2ml water to dissolve it with shaking and the solution show green color. Write out the equation of the reaction.

4.2 Disproportionation of K_2MnO_4

Add dilute acetic acid to a little K_2MnO_4 solution prepared above, observe the change of the solution color and the formation of precipitate. Write out ion reaction equation.

5. Mn(Ⅶ) compounds

Take three test tubes, add a drop of 0.1mol/L $KMnO_4$ solution in each one. Add 5 drops of 1mol/L H_2SO_4, distilled water and 6mol/L NaOH to the three tubes respectively. Then add a drop of 0.1mol/L Na_2SO_3 to each tube and observe the phenomenon. Write out the reaction equation and make a conclusion about the effects of media on the reduction product of $KMnO_4$.

6. Fe (Ⅱ) compounds

Take one test tube, add 2ml distilled water and 1~2 drops of 2mol/L H_2SO_4, then add some ferrous ammonium sulfate crystal. Add 1ml 2mol/L boiling NaOH in another test tube to the ferrous ammonium sulfate solution rapidly (without shaken), observe the phenomenon. Then shake it and place for a moment, observe the color change of the precipitate, and explain the reasons of every procedures and the change of phenomenon. Write out the ion equation of reaction.

7. Fe (Ⅲ) compounds

7.1 Add 3 drops of 2mol/L NaOH to 4 drops of 0.1mol/L $FeCl_3$ solution, observe the phenomenon and write out the reaction equation.

7.2 Add 2 drops of 0.1mol/L KI to 4 drops of 0.1mol/L $FeCl_3$ solution, observe the phenomenon and try to check out the products. Write out ion equation of the reaction.

8. Complexes of Fe (Ⅱ), Fe (Ⅲ)

8.1 Add a drop of 0.1mol/L $K_4[Fe(CN)_6]$ to 5 drops of 0.1mol/L $FeCl_3$ solution, observe the formation of prussian blue precipitate (or sol).

8.2 Add a drop of 0.1mol/L $K_3[Fe(CN)_6]$ to 5 drops of 0.1mol/L $(NH_4)_2Fe(SO_4)_2$ solution, observe the formation of Tengs blue precipitate (or sol).

8.3 Add a drop of 2mol/L H_2SO_4 and 0.1mol/L KSCN solution to 5 drops of 0.1mol/L $(NH_4)_2Fe(SO_4)_2$ solution, observe whether the color changes. Then add a drop of 3% H_2O_2 solution, observe the color changes and write out the ion reaction equation.

Notes

1. When testing the reducibility of Cr^{3+}, and H_2O_2 being used as the oxidant, the solution will appear brownish red sometimes, which is due to the formation of sodium perchromate.

2. In acidic solution, when MnO_4^- is reduced to Mn^{2+}, brown precipitate of MnO_2 appears sometimes, this is because the concentration of reducing agent is not enough, and the product Mn^{2+} reacts with excessive MnO_4^-:

$$2 MnO_4^- + 3Mn^{2+} + 2H_2O = 5MnO_2\downarrow + 4H^+$$

Questions

(1) Preview requirements

(a) Review the important properties of various main compounds of chromium and manganese in inorganic chemistry textbooks, and focus on clarifying the conditions for the conversion of various valence states.

(b) Review the contents of iron in textbooks, and focus on the changing law of the stability of +2 and +3 oxidation states and the conditions of mutual transformation, as well as the properties and important reactions of their coordination compounds.

(2) Thinking Questions

(a) How to identify the presence of Cr^{3+} or Mn^{2+}?

(b) How to store $KMnO_4$ solution? Why?

(c) Try to use two methods to realize the mutual transformation of Fe^{2+} and Fe^{3+}.

实验十　药用氯化钠的制备

实验目的

1. **掌握**　药用氯化钠的制备原理和方法。溶解、过滤、沉淀、蒸发、结晶等基本操作。
2. **了解**　沉淀－溶解平衡的原理和应用。

实验原理

粗食盐中含有挥发性有机杂质、可溶性无机杂质（K^+、Ca^{2+}、Mg^{2+}、Fe^{3+}、SO_4^{2-}、Br^-、I^- 和重金属）及泥沙等机械杂质。这些杂质的去除方法如下。

1. 有机杂质可采用炒的方法去除。
2. 不溶性机械杂质如泥沙等可采用过滤法去除。
3. 一些可溶性的杂质可根据其性质借助于化学法除去。如加入 $BaCl_2$ 溶液可使 SO_4^{2-} 离子转化为 $BaSO_4$ 沉淀，加入 $NaOH$ 和 Na_2CO_3 的混合溶液可使 Ca^{2+}、Mg^{2+}、Fe^{3+} 和 Ba^{2+} 等离子生成难溶物沉淀，先后过滤除去。最后再用盐酸将溶液调至微酸性，除去过量的 $NaOH$ 和 Na_2CO_3。

$$Ba^{2+}+SO_4^{2-} \rightleftharpoons BaSO_4 \downarrow$$
$$Ca^{2+}+CO_3^{2-} \rightleftharpoons CaCO_3 \downarrow$$
$$2Mg^{2+}+2OH^-+CO_3^{2-} \rightleftharpoons Mg_2(OH)_2CO_3 \downarrow$$
$$Ba^{2+}+CO_3^{2-} \rightleftharpoons BaCO_3 \downarrow$$
$$Fe^{2+}+3OH^- \rightleftharpoons Fe(OH)_3 \downarrow$$

4. 少量的可溶性杂质如 Br^-、I^-、K^+ 等离子，可根据溶解度的不同，在重结晶时，使其残留在母液中而弃去。

仪器、试剂及其他

1. **仪器**　电磁炉，烧杯，量筒，试管，布氏漏斗，蒸发皿，台秤，抽滤装置。
2. **试剂**　粗食盐，2mol/L HCl，饱和 H_2S 溶液，6mol/L NaOH，饱和 Na_2CO_3 溶液，25% $BaCl_2$，1mol/L $BaCl_2$。
3. **其他**　pH 试纸，滤纸。

实验内容

1. **有机及机械杂质的去除**　称取 20.0g 粗食盐，小火炒至有机物炭化，转移至 250ml 烧杯中，加入 70ml 水，加热搅拌使粗食盐完全溶解，趁热倾泻法过滤，得滤液。

2. **SO_4^{2-} 的去除**　滤液加热近沸，边搅拌边滴加入 25% $BaCl_2$ 溶液（约 4ml 左右）至 SO_4^{2-} 沉淀完全。继续加热 5 分钟，趁热过滤，弃去沉淀。

3. **重金属及 Mg^{2+}、Ca^{2+}、Ba^{2+}、Fe^{3+} 等离子的去除**　在滤液中加入饱和硫化氢溶液数滴，若无沉淀，不必再加。继续加入 6mol/L NaOH 与饱和 Na_2CO_3 混合溶液（体积比 1∶1）约 10ml，将溶液 pH 调至 10~11，待沉淀完全后，加热至微沸，静置冷却，过滤，弃去沉淀。

4. 过量 CO_3^{2-}、OH^- 的去除　滤液转移至蒸发皿中，滴加 2mol/L HCl 溶液，直至溶液 pH 为 3~4。

5. 浓缩与结晶　将滤液倒入蒸发皿中，蒸发浓缩至糊状，有大量 NaCl 结晶出现（切不可蒸干），趁热减压抽滤，将所得产品转移到蒸发皿中，小火烘（炒）干。将干燥冷却后的氯化钠晶体称量，计算产率。

所得氯化钠晶体装入袋中，贴上标签，以供杂质限度检查。

注意事项

1. 沉淀剂用量要过量，滴加后再应煮沸几分钟，使沉淀由小颗粒聚沉为大颗粒，以利于沉淀与溶液的分离。
2. 检查沉淀是否完全的方法：可吸取少量清液到试管中，加 1~2 滴沉淀剂，无浑浊即表示已沉淀完全。
3. 浓缩 NaCl 溶液时用小火加热，保持溶液微沸并不停搅拌，切不可蒸干。
4. 为防止炒干后的 NaCl 结成块状，炒干时应小火加热且不断搅拌。

思考题

1. 在除去化学杂质时，为什么要先加入 $BaCl_2$ 溶液，然后再加入 Na_2CO_3 溶液，最后加 HCl 溶液？能否改变加入沉淀剂的次序？
2. 如何除去过量的沉淀剂 $BaCl_2$、$NaSO_4$ 和 NaOH？
3. 在浓缩过程中，能否把溶液蒸干？为什么？

Experiment 10 Preparation of Medicinal Sodium Chloride

Objectives

1. Master principle and method of how to prepare medicinal sodium chloride. To master basic operations such as dissolution, filtration, precipitation, evaporation and crystallization.

2. Understand the principle and application of sediment-solution equilibrium.

Principle

Coarse salt contains volatile organic impurities, soluble inorganic impurities (K^+, Ca^{2+}, Mg^{2+}, Fe^{3+}, SO_4^{2-}, Br^-, I^- and heavy metals etc.), and mechanical impurities such as sediment. The methods for the removal of these impurities are as follows.

1. The organic impurities can be removed by the sauté method.

2. The insoluble impurities such as sediment are removed by filtration.

3. Some soluble impurities can be removed by precipitation basing on their chemical properties. For example, sulfate can be separated as $BaSO_4$ by solution of $BaCl_2$; Ca^{2+}, Mg^{2+}, Fe^{3+}, Ba^{2+} etc. can be removed as insoluble precipitates by the mixture of NaOH and Na_2CO_3. Finally, the solution was adjusted to slightly acidic with hydrochloric acid to remove the excessive NaOH and Na_2CO_3.

$$Ba^{2+} + SO_4^{2-} \rightleftharpoons BaSO_4 \downarrow$$
$$Ca^{2+} + CO_3^{2-} \rightleftharpoons CaCO_3 \downarrow$$
$$2Mg^{2+} + 2OH^- + CO_3^{2-} \rightleftharpoons Mg_2(OH)_2CO_3 \downarrow$$
$$Ba^{2+} + CO_3^{2-} \rightleftharpoons BaCO_3 \downarrow$$
$$Fe^{2+} + 3OH^- \rightleftharpoons Fe(OH)_3 \downarrow$$

4. Small amount of soluble impurities such as Br^-, I^- and K^+, having different solubility in sodium chloride, can be left in the mother liquor and discarded during recrystallization.

Instruments, Chemicals and Others

1. Equipment

Induction cooker, beaker, measuring cylinder, test tube, buchner funnel, evaporating dish, platform balance, the suction filter device.

2. Reagents

Coarse salt, 2mol/L HCl, saturated hydrogen sulfide solution, 6mol/L NaOH, saturated sodium carbonate solution, 25% $BaCl_2$, 1mol/L $BaCl_2$.

3. Other

pH test paper, filter paper.

Experimental Procedures

1. Remove of organic impurities and mechanical impurities

Weigh 20.0g coarse salt, heat moderately to carbonize the organic impurities, and then dissolve the solid in a beaker (250ml, 70ml water) while heating. Finally, filter it with suction by tilt-pour process while hot. Retain the filtrate.

2. Removal of SO_4^{2-}

Heat the filtrate to keep the solution on the simmer, and add 25% $BaCl_2$ solution (about 4ml) while stirring until the sulfate precipitates completely. Keep heated for 5min, and then filter with suction by tilt-pour process while hot. Discard the filter residue.

3. Removal of heavy metal ion, Mg^{2+}, Ca^{2+}, Ba^{2+} and Fe^{3+}

Add a few drops of saturated hydrogen sulfide solution to the filtrate. Add the mixture of 6mol/L NaOH and saturated Na_2CO_3 (volume ratio 1:1, 10ml) to adjust the pH to 10~11. Then, boil and cool down, filter by tilt-pour process. Retain the filtrate.

4. Removal of excessive CO_3^{2-}, OH^-

Transfer the filtrate to the evaporating dish, and neutralize with 2mol/L HCl solution until the pH 3~4.

5. Concentration and crystallization

Evaporate the solution to be thick paste, a lot of crystal precipitate. Filter with vacuum suction while hot. Transfer the obtained product to the evaporating dish. Dry and then cool down. Weigh and calculate the yield of the sodium chloride crystals.

The sodium chloride crystals were packaged and labeled for the examination of impurities' limitation.

Notes

1. The amount of precipitants added in the solution should be slightly excessive, and the solution should be boiled for several minutes after dripping, so that the small particles could precipitate to large particles to facilitate the separation of precipitation and solution.

2. A method used to check the completeness of the precipitate: absorb a small amount of supernatant into the test tube and add 1~2 drops of precipitant, no turbidity means that the precipitation has been complete.

3. During concentration, NaCl solution should be moderately heated while keeping the solution slightly boiling and stirring. It shouldn't be evaporated to dryness.

4. Moderately heat and keep stirring to prevent NaCl from forming a block during the drying process.

Questions

(1) Why, during the purification of NaCl, should the agents be added sequentially $BaCl_2$, Na_2CO_3, and HCl? Can we change the order of precipitants?

(2) How to remove the excess precipitants: $BaCl_2$, Na_2SO_4 and NaOH?

(3) Can we evaporate the condensed solution to dryness? Why?

实验十一 药用氯化钠的性质及杂质限量的检查

实验目的

了解 《中国药典（2020年版）》对药用氯化钠的鉴别、检查方法。

实验原理

1. 鉴别实验是被检药品组成或离子的特征试验，即氯化钠的组成离子 Na^+ 和 Cl^-。

2. 钡盐、钾盐、钙盐、镁盐及硫酸盐的限度检验，是根据沉淀反应的原理，样品管和标准管在相同条件下进行比浊试验，样品管不得比标准管更深。

3. 重金属系 Pb、Bi、Cu、Hg、Sb、Sn、Co、Zn 等金属离子，它们在一定条件下能与 H_2S 或 Na_2S 作用而显色。《中国药典》规定，在弱酸条件下进行，用稀醋酸调节。实验证明，在 pH=3 时，PbS 沉淀最完全。

重金属的检查，是在相同条件下进行比色试验。

仪器、试剂及其他

1. **仪器** 蒸发皿，烧杯，漏斗，抽滤瓶，奈氏比色管，离心机。

2. **试剂**

酸：饱和 H_2S 溶液，0.1mol/L HCl，2mol/L HCl，0.5mol/L H_2SO_4，0.1mol/L HAc，3mol/L HAc。

碱：$NH_3 \cdot H_2O$。

盐：25% $BaCl_2$，饱和 Na_2CO_3，0.1mol/L $AgNO_3$，0.1mol/L $KMnO_4$，0.1mol/L KI，0.1mol/L KBr，0.1mol/L $(NH_4)_2S_2O_8$，0.1mol/L NH_4SCN，四苯硼钠溶液，0.1mol/L Na_2HPO_4，0.1mol/L $(NH_4)_2C_2O_4$，0.1mol/L $CaCl_2$。

3. **其他** 药用氯化钠（自己制备），淀粉KI试纸，三氯甲烷，氯水，标准硫酸钾溶液，标准铁盐溶液，标准铅盐溶液。

实验步骤

1. **氯化物的鉴别反应**

（1）生成氯化银沉淀 取产品少许溶液，加硝酸银溶液，即生成白色凝乳状沉淀，沉淀溶于氨试液，但不溶于硝酸。

$$Cl^- + Ag^+ \rightarrow AgCl\downarrow$$

（2）还原性实验 取产品少许，加水溶解后，加 $KMnO_4$ 与稀 H_2SO_4 加热，即产生氯气，遇淀粉KI试纸即显蓝色。

$$10Cl^- + 2MnO_4^- + 16H^+ \rightarrow 5Cl_2\uparrow + 2Mn^{2+} + 8H_2O$$

2. **碘化物与溴化物** 取产品 2g，加蒸馏水 6ml 溶解后，加氯仿 1ml，并用等量蒸馏水稀释的氯水试液，随滴随振摇，氯仿层不得显紫红色、黄色或橙色。

对照试验：分别取碘化物和溴化物溶液各 1ml，分置于 2 支试管内，各加氯仿 0.5ml，并滴加氯试液，振摇。两试管中分别显示紫红色、黄色或红棕色。

$$2Br^- + Cl_2 \rightarrow Br_2 + 2Cl^-$$
$$2I^- + Cl_2 \rightarrow I_2 + 2Cl^-$$

3. 钡盐 取产品 4g，用蒸馏水 20ml 溶解，过滤，滤液分为两等份，一份中加稀 H_2SO_4 2ml，静置 2 小时，两液应同样透明。

4. 钾盐 取产品 5.0g，加水 20ml 溶解后，加稀醋酸 2 滴，加四苯硼钠溶液（取四苯硼钠 1.5g，置乳钵中，加水 10ml 后研磨，再加水 40ml，研匀，用质密的滤纸滤过，即得）2ml，加水使成 50ml，显浑浊，与标准硫酸钾溶液 12.3ml 用同一方法制成的对照液比较，不得更浓（0.02%），反应式为：

$$K^+ + B(C_6H_5)_4^- \rightarrow KB(C_6H_5)_4 \downarrow（白色）$$

标准硫酸钾溶液的制备：精密称取在 105℃干燥至恒重的硫酸钾 0.181g，置 1000ml 量瓶中，加水适量使溶解并稀释至刻度，摇匀，即得（每 1ml 相当于 81.1μg 钾）。

5. 硫酸盐 取 50ml 奈氏比色管两支，A 管中加标准硫酸钾溶液 1ml（每 1ml 标准硫酸钾溶液相当于 100μg 的 SO_4^{2-}），加蒸馏水稀释至约 25ml 后，加 0.1mol/L HCl 1ml，置 30~35℃水浴中，保温 10 分钟，加 25% $BaCl_2$ 溶液 3ml，加适量水使成 50ml，摇匀，放置 10 分钟。

取产品 5g 置 B 管中，加水溶解至约 25ml，溶液应透明，如不透明可过滤，于滤液中加 0.1mol/L HCl 1ml，置 30~35℃水浴中，保温 10 分钟。加 25% $BaCl_2$ 溶液 3ml，用蒸馏水稀释，使成 50ml，摇匀，放置 10 分钟。

A、B 两管放置 10 分钟后，置比色架上，在光线明亮处双眼由上而下透视，比较两管的浑浊度，乙管发生的浑浊度不得高于甲管 (0.002%)。

6. 钙盐与镁盐 取产品 4g，加水 20ml 溶解后，加氨试液 2ml 摇匀，分成两等份。一份加草酸铵试液 1ml，另一份加磷酸氢二钠试液 1ml，5 分钟内均不得发生浑浊。

对比试验如下。

（1）取钙盐溶液 1ml，加草酸铵试液 1ml，滴加氨水至显微碱性，溶液有白色结晶析出。反应式为：

$$Ca^{2+} + C_2O_4^{2-} \rightarrow CaC_2O_4 \downarrow（白色）$$

（2）取镁盐溶液 1ml，加磷酸氢二钠 1ml，加氨水 10 滴，有白色结晶析出。反应式为：

$$Mg^{2+} + HPO_4^{2-} + NH_3·H_2O \rightarrow MgNH_4PO_4 \downarrow（白色）+ H_2O$$

7. 铁盐 取产品 5g，置于 50ml 奈氏比色管中，加蒸馏水 35ml 溶解，加 0.1mol/L HCl 5mL，新配 0.1mol/L 过硫酸铵几滴，再加硫氢化铵试液 5ml、适量蒸馏水使成 50ml，摇匀。如显色与标准溶液 1.5ml（用同法处理后制得的标注管）颜色比较，不得更深（0.0003%）。

标准铁盐溶液的制备：精密称取未风化的硫酸铁铵 0.8630g，溶解后转入 1000ml 容量瓶中，加硫酸 2.5ml，加水稀释至刻度，摇匀。临用时精密量取 10ml，置于 100ml 的量瓶中，加水稀释至刻度，摇匀，即得每 1ml 相当于 10μg 的 Fe。

8. 重金属 取 50ml 奈氏比色管两支，A 管中加标准铅溶液 (10μg Pb/ml) 1ml，加稀醋酸 2ml，加水稀释至 25ml。于 B 管中加产品 5g，加水 20ml 溶解后，加稀醋酸 2ml 与水适量使成 25ml。A、B 两管中分别加硫化氢试液各 10ml，摇匀，在暗处放置 10 分钟，同置白纸上，自上面透视，B 管中显出的颜色与 A 管比较，不得更深 (含重金属不得超过百分之二)。

铅储备液的制备：精密称取在 105℃干燥至恒重的硝酸铅 0.1598g，加硝酸 5ml 与水 50ml，溶解后，按规定配制成 1000ml，摇匀，即得（每 1ml 相当于 8100μg 的 Pb)。

标准铅溶液的制备：精密量取铅储备液 10ml，置 100ml 容量瓶中，加水稀释至刻度，摇匀，即得（每 1ml 相当于 10μg 的 Pb)。

重金属检测中,标准铅溶液应新鲜配制。配制与存用的玻璃容器均不得含有铅。

1. 本实验中鉴别反应的原理是什么?
2. 何种离子的检验可选用比色试验?何种分析方法称为限量分析?

Experiment 11 Inspection of Properties and Impurity Limits of Medicinal Sodium Chloride

Objectives

Understand the identification and examination methods of medicinal sodium chloride in Chinese Pharmacopoeia (2020 edition).

Principle

1. The identification test is the characteristic test of the composition or ion of the tested drug, that is, the composition ion of Na^+ and Cl^-.

2. The limit test of barium salt, potassium salt, calcium salt, magnesium salt and sulfate is based on the principle of precipitation reaction. The turbidimetric test of sample tube and standard tube shall be carried out under the same conditions, and the sample tube shall not be deeper than the standard tube.

3. Heavy metals, such as Pb, Bi, Cu, Hg, Sb, Sn, Co, Zn metal ions, can react with H_2S or Na_2S and develop color under certain conditions. According to Chinese Pharmacopoeia, it is regulated by dilute acetic acid under the condition of weak acid. The results show that the precipitation of PbS is the most complete at pH=3.

The inspection of heavy metals is a colorimetric test under the same conditions.

Equipment, Chemicals and Others

1. Equipment

Evaporating dish, beaker, funnel, suction flask, Nessler colorimetric tube, centrifuge.

2. Chemicals

Acid: saturated H_2S solution, 0.1mol/L HCl, 2mol/L HCl, 0.5mol/L H_2SO_4, 0.1mol/L HAc, 3mol/L HAc.

Alkali: $NH_3 \cdot H_2O$.

Salt: 25% $BaCl_2$, saturated Na_2CO_3 solution, 0.1mol/L $AgNO_3$, 0.1mol/L $KMnO_4$, 0.1mol/L KI, 0.1mol/L KBr, 0.1mol/L $(NH_4)_2S_2O_8$, 0.1mol/L NH_4SCN, sodium tetraphenylboron solution, 0.1mol/L Na_2HPO_4, 0.1mol/L $(NH_4)_2C_2O_4$, 0.1mol/L $CaCl_2$.

3. Others

Self-made medicinal sodium chloride, KI starch test paper, chloroform, chlorine water, standard potassium sulfate solution, standard iron salt solution and standard lead salt solution.

Experimental Procedures

1. Identification Reaction of Chloride

1.1 Generation of Silver Chloride Precipitation

Take a little solution of the product and add silver nitrate solution to form a white coagulated precipitate, which is soluble in ammonia solution but insoluble in nitric acid.

$$Cl^- + Ag^+ \rightarrow AgCl \downarrow$$

1.2 Reducibility Experiment

Take a little of the product, add water to dissolve it. Add $KMnO_4$ and dilute sulphuric acid, heat it to produce chlorine gas, which will appear blue in case of KI starch test paper.

$$10Cl^- + 2MnO_4^- + 16H^+ \rightarrow 5Cl_2 \uparrow + 2Mn^{2+} + 8H_2O$$

2. Iodide and Bromide

Take 2g of the product, add 6ml of distilled water to dissolve it, add 1ml of chloroform, and add the chlorine water solution diluted with the same amount of distilled water. Drop chlorine test solution, shake constantly. The chloroform layer shall not show purplish red, yellow or reddish brown in the test tube.

Control test: take 1ml of iodide and bromide solution respectively, put them into two test tubes, add 0.5ml of chloroform each, drop the chlorine solution and then shake it. The two tubes showed purplish red, yellow or reddish brown respectively.

$$2Br^- + Cl_2 \rightarrow Br_2 + 2Cl^-$$
$$2I^- + Cl_2 \rightarrow I_2 + 2Cl^-$$

3. Barium Salt

Take 4g of the product, dissolve it with 20ml of distilled water, then filter it. Divide the filtrate solution into two equal parts, add 2ml of dilute H_2SO_4 into one part, and leave it for 2 hours, the two solutions should be equally transparent.

4. Sylvite

Take 5.0g of the product, add 20ml of water to dissolve it, add 2 drops of dilute acetic acid, add 2ml of sodium tetraphenylborate solution (take 1.5g of sodium tetraphenylborate, put it in a mortar, add 10ml of water to grind it, then add 40ml of water, grind it well, filter it with dense filter paper to get the solution). Add water to make it 50ml solution, appearing turbid. Compare it with 12.3ml of standard potassium sulfate solution with the same method, no more concentrated (0.02%). The reaction formula is:

$$K^+ + B(C_6H_5)_4^- \rightarrow KB(C_6H_5)_4 \downarrow (White)$$

Preparation of standard potassium sulphate solution: weigh precisely 0.181g of potassium sulphate which is dried to constant weight at 105℃. Put it into a 1000ml measuring bottle, add some water to dissolve and dilute it to the scale, shake it well, and then get the solution (1ml solution is equivalent to 81.1μg of potassium).

5. Sulphate

Take two 50ml Nessler colorimetric tubes, add 1ml of standard potassium sulphate solution (each 1ml of standard potassium sulphate solution is equivalent to 100μg SO_4^{2-}) into tube A, dilute to about 25ml with distilled water, add 1mL 0.1mol/L HCl, place in a 30~35℃ water bath, keep warm for 10 minutes. Add 3ml of 25% $BaCl_2$ solution, add appropriate water to make 50ml solution, shake well, and place for 10 minutes.

Take 5g of this product and put it into tube B, add water to dissolve it to about 25ml, the solution should be transparent. If not, it can be filtered. Add 1ml of 0.1mol/L HCl into the filtrate solution, put it into a water bath at 30~35℃, and keep it warm for 10 minutes. Add 3ml of 25% $BaCl_2$ solution, dilute with distilled water to 50ml, shake well, and place for 10min.

After 10 minutes of placement, place tube A and tube B on the color comparison frame, and observe them from top to bottom in the bright light place. Compare the turbidity of the two tubes. The turbidity of tube B shall not be higher than that of tube A (0.002%).

6. Calcium Salt and Magnesium Salt

Take 4g of this product, add 20ml of water to dissolve, add 2ml of ammonia solution, shake well, and divide into two equal parts. Add 1ml ammonium oxalate solution into one part and 1ml disodium hydrogen phosphate into another, test solution shall not be turbid within 5 minutes.

Comparative test:

(1) Take 1ml of calcium salt solution, add 1ml of ammonium oxalate solution, drop ammonia water to micro alkaline, the solution produces white crystallization. The reaction formula is:
$$Ca^{2+} + C_2O_4^{2-} \rightarrow CaC_2O_4 \downarrow \text{ (White)}$$

(2) Take 1ml of magnesium salt solution, 1ml of disodium hydrogen phosphate solution, add 10 drops of ammonia water, there is white crystallization. The reaction formula is:
$$Mg^{2+} + HPO_4^{2-} + NH_3 \cdot H_2O \rightarrow MgNH_4PO_4 \downarrow \text{ (White) } + H_2O$$

7. Iron Salt

Take 5g of this product, put it into 50ml Nessler colorimetric tube, add 35ml of distilled water to dissolve it, add 5ml 0.1mol/L HCl, add a few drops of newly prepared 0.1mol/L ammonium persulfate, add 5ml of ammonium bisulfide solution and a proper amount of distilled water to make it 50ml, shake it. The color of the test tube compared with the tube of 1.5ml standard iron salt solution prepared by the same method shall not be deeper (0.0003%).

Preparation of the standard iron salt solution: weigh precisely 0.8630g of ammonium ferric sulfate, dissolve it, transfers it into a 1000ml volumetric flask. Add 2.5ml of sulfuric acid, add water to the scale, and shake well. When in use, put 10ml of the solution into a 100ml measuring bottle, dilute it to the scale, and shake it well, every 1ml of the solution is equivalent to 10μg Fe.

8. Heavy Metals

Take two 50ml Nessler colorimetric tubes, add 1ml of standard lead solution (10μg Pb/ml) and 2ml of dilute acetic acid into tube A, dilute to 25ml with water. Add 5g of sample into tube B, add 20ml of water to dissolve, add 2ml of dilute acetic acid and proper amount of water to make 25ml solution. Add 10ml of hydrogen sulfide solution to each of the two tubes of A and B respectively, shake well, place them in a dark place for 10 minutes. Place the tubes on white paper at the same time, perspective from above, the color shown in tube B shall not be deeper than that of tube A (the content of heavy metal shall not exceed 2%).

Preparation of lead stock solution: weigh precisely 0.1598g of lead nitrate dried at 105℃ to constant weight, add 5ml of nitric acid and 50ml of water, dissolve it to 1000ml as required, shake it well (1ml solution is equivalent to 8100μg Pb).

Preparation of standard lead solution: accurately measure 10ml of lead stock solution, place it in a 100ml volumetric flask, add water to dilute it to the scale, shake it well (1ml is equivalent to 10μg Pb).

Notes

The standard lead solution used in heavy metals detection shall be freshly prepared. The glass containers for preparation and storage solution shall not contain lead.

Questions

(1) What is the principle of identification reaction in this experiment?

(2) Which ions can be tested by colorimetric test? Which analytical method is called limit analysis?

实验十二　硫酸亚铁铵的制备

实验目的

1. 了解　复盐硫酸亚铁铵的制备方法。
2. 练习和巩固水浴加热、蒸发浓缩、结晶、减压过滤等基本操作。
3. 学习用目视比色法检验产品的质量等级。

实验原理

硫酸亚铁铵 $(NH_4)_2Fe(SO_4)_2·6H_2O$，又称摩尔盐，是浅蓝绿色单斜晶体，易溶于水，难溶于乙醇。在空气中比亚铁盐稳定，不易被氧化，在定量分析中常用于来配制亚铁离子的标准溶液。

常用的制备方法是先用铁与稀硫酸作用制得 $FeSO_4$，$FeSO_4$ 再与等量的 $(NH_4)_2SO_4$ 在水溶液中生成硫酸亚铁铵。由于复盐在水中的溶解度比组成它的每一组分的溶解度都小，因此溶液经蒸发浓缩、冷却后，复盐在水溶液中首先结晶，形成 $(NH_4)_2Fe(SO_4)_2·6H_2O$ 晶体。

$$Fe+H_2SO_4=FeSO_4+H_2\uparrow$$
$$FeSO_4+(NH_4)_2SO_4+6H_2O=(NH_4)_2Fe(SO_4)_2·6H_2O$$

产品中主要的杂质是 Fe^{3+}，产品质量的等级也常以 Fe^{3+} 含量多少来衡量。产品中 Fe^{3+} 的含量可用目视比色法来测定。

将样品配制成溶液，在一定条件下，与含一定量杂质离子的系列标准溶液进行比色或比浊，以确定杂质含量范围。如果样品溶液的颜色或浊度不深于标准溶液，则认为杂质含量低于某一规定限度，这种分析方法称为限量分析。

仪器、试剂及其他

1. **仪器**　电子天平，锥形瓶，烧杯，量筒，移液管，奈氏比色管，玻璃棒，漏斗，布氏漏斗，抽滤瓶，表面皿，蒸发皿，真空泵，酒精灯，温度计，电炉，石棉网，铁架台，铁圈，水浴锅。
2. **试剂**　3mol/L H_2SO_4，40% NaOH，25% KSCN，0.1mol/L $K_3[Fe(CN)_6]$，25% $BaCl_2$，$(NH_4)_2SO_4(s)$，铁屑，95% 乙醇。
3. **其他**　滤纸，pH试纸。

实验步骤

1. **铁屑的净化（除去油污）**　用电子天平称取 2.0g 铁屑，放入 150ml 锥形瓶中，加入 20ml 10% Na_2CO_3 溶液，加热煮沸除去油污。倾去碱液，用水洗铁屑至中性（如果用纯净的铁屑，可省去这一步）。

2. **$FeSO_4$ 溶液的制备**　往盛有铁屑的锥形瓶中加入 15ml 3mol/L 的 H_2SO_4 溶液，置于通风橱中水浴加热（温度低于80℃）至不再有气体冒出为止（该反应约需40分钟，此时可配制硫酸铵饱和溶液）。在加热过程中需适当补充些热水，以保持原体积。反应完全后敞开约2分钟，以除去一些有毒气体。趁热抽滤，滤液迅速转移至盛有饱和 $(NH_4)_2SO_4$ 溶液的蒸发皿中。用少量热水洗涤锥形瓶及残渣。收集残渣，用滤纸吸干后称量，根据已反应掉的铁屑质量，算出理论上溶液中 $FeSO_4$ 的含量。

3. 硫酸铵饱和溶液的配制　根据硫酸亚铁的理论产量计算所需硫酸铵的质量，参照表 3-4 称取相应量将其配成饱和溶液。

表 3-4　硫酸铵的溶解度（单位：$g/100g\ H_2O$）

温度 /℃	10	20	30	40	50
溶解度	70.6	73.0	75.4	78.0	81.0

4. 硫酸亚铁铵的制备　将混合液在酒精灯或电炉上蒸发，浓缩至溶液表面刚出现结晶膜时为止。静置冷却至室温，即有硫酸亚铁铵晶体析出，观察晶体颜色。用布氏漏斗抽滤，尽可能使母液与晶体分离完全，用少量乙醇洗涤晶体。将晶体取出，置于两张洁净的滤纸之间，并轻压以吸干母液；称量，计算产量和产率。

5. Fe^{3+} 的质量检测（限量分析）

（1）配制浓度为 0.0100 mg/ml 的 Fe^{3+} 标准溶液　准确称取 0.0216g $(NH_4)Fe(SO_4)_2·6H_2O$ 于烧杯中，先加入少量蒸馏水溶解，再加入 6ml 的 3mol/L H_2SO_4 溶液酸化，用蒸馏水将溶液在 250ml 容量瓶中定容。此溶液中 Fe^{3+} 浓度即为 0.0100mg/ml。

（2）配制标准色阶　用移液管分别移取 Fe^{3+} 标准溶液 5.00、10.00、20.00ml 于比色管中，各加 1ml 3mol/L 的 H_2SO_4 和 1ml 25% 的 KSCN 溶液，再用新煮沸过放冷的蒸馏水将溶液稀释至 25ml，摇匀，即得到含 Fe^{3+} 量分别为 0.05mg（一级）、0.10mg（二级）和 0.20mg（三级）的三个等级试剂标准液。

（3）产品等级的确定　称取 1.0g 硫酸亚铁铵晶体，加入 15ml 不含氧的蒸馏水溶解，定量转移至 25ml 比色管中，再加 1ml 3mol/L H_2SO_4 和 1ml 25% KSCN 溶液，最后加入不含氧的蒸馏水将溶液稀释到 25ml，摇匀，与标准溶液进行目视比色，确定产品的等级。

注意事项

1. 在制备 $FeSO_4$ 时，水浴温度不要超过 80℃，以免反应过于剧烈。
2. 在制备 $FeSO_4$ 时，保持溶液 pH≤1，以使铁屑与硫酸溶液的反应能不断进行。
3. 在检验产品中 Fe^{3+} 含量时，为防止 Fe^{3+} 被溶解在水中的氧气氧化，可将蒸馏水加热至沸腾，以去除水中溶入的氧气。
4. 制备硫酸亚铁铵晶体时，溶液必须呈酸性，蒸发浓缩时不需要搅拌、不可浓缩至干。

思考题

1. 水浴加热时应注意什么问题？
2. 怎样确定所需要的硫酸铵用量？如何配制硫酸铵饱和溶液？
3. 为什么在制备硫酸亚铁铵时要使铁过量？
4. 为什么制备硫酸亚铁铵时要保持溶液有较强的酸性？

Experiment 12 Preparation of Ammonium Ferrous Sulfate

Objectives

1. Understand the preparation method of $(NH_4)_2Fe(SO_4)_2 \cdot 6H_2O$.
2. Practice and consolidate the basic operations of water bath heating, evaporating, concentrating, crystallizing and vacuum filtrating.
3. Learn the technology of product quality analysis by visual colorimetry.

Principles

Ammonium ferrous sulfate is also named molar salt, $(NH_4)_2Fe(SO_4)_2 \cdot 6H_2O$, which is a kind of transparent bluish-green monoclinic crystal, easily soluble in water and hardly soluble in ethanol. It is more stable than ferrous salt, and not to be oxidized in air. In quantitative analysis, it is often used to prepare ferrous ion standard solution.

Common preparation method is to dissolve scrap iron in dilute H_2SO_4. Thus $FeSO_4$ solution is prepared. Then $FeSO_4$ reacts with $(NH_4)_2SO_4$ of the same amount to form ammonium ferrous sulfate in aqueous solution, because the solubility of double salt is smaller than its component, so crystal of ammonium ferrous sulfate with small solubility forms after evaporation, concentration and cooling.

$$Fe+H_2SO_4=FeSO_4+H_2\uparrow$$
$$FeSO_4+(NH_4)_2SO_4+6H_2O=(NH_4)_2Fe(SO_4)_2 \cdot 6H_2O$$

The main impurity in the product is Fe^{3+}, and the quality level of the product is often measured by the content of Fe^{3+}. The content of Fe^{3+} in the product can be determined by visual colorimetry.

Prepare the sample into a solution, under certain conditions, compare with a series of standard solutions containing a certain amount of impurity ions for color or turbidity, so as to determine the impurity content range. If the color or turbidity of the sample solution is not deeper than that of the standard solution, it is considered that the impurity content is lower than a specified limit. This analysis method is called limit analysis.

Equipment, Chemicals and Others

1. Equipment
Electronic balance, conical flask, beaker, measuring cylinder, pipette, Nessler colorimetric tube, glass rod, funnel, Buchner funnel, suction filter bottle, watch glass, evaporating dish, vacuum pump, alcohol burner, thermometer, electric furnace, asbestos gauze, iron stand, iron ring, water bath pot.

2. Chemicals
3mol/L H_2SO_4, 40% NaOH, 25% KSCN, 0.1mol/L $K_3[Fe(CN)_6]$, 25% $BaCl_2$, $(NH_4)_2SO_4$(s), scrap iron, 95% ethanol.

3. Others

Filter paper, pH test paper.

1. Purification of Iron Scraps (Oil Removal)

Weigh 2.0g of iron scraps with electronic balance, put them into a 150ml conical flask, add 20ml of 10% Na_2CO_3 solution, heat and boil to remove oil stains. Remove the alkali solution and wash the iron scraps with water to neutral (if pure iron scraps are used, this step can be omitted).

2. Preparation of $FeSO_4$ Solution

Add 15ml of 3mol/L H_2SO_4 solution into the conical flask containing iron scraps, heat it in a water bath in a fume hood (the temperature is lower than 80℃) until no gas is emitted (the reaction takes about 40 minutes, at this time, ammonium sulfate saturated solution can be prepared). During the heating process, it is necessary to add some hot water to maintain the original volume. Open for about 2 minutes after the reaction to remove some toxic gases. Filter while hot, the filtrate is transferred to the evaporating dish containing saturated $(NH_4)_2SO_4$ solution. Wash conical flask and residue with a small amount of hot water. Collect the residue, dry it with filter paper, weigh it, and calculate the content of $FeSO_4$ in the solution theoretically according to the mass of iron scraps that have been reacted.

3. Preparation of Saturated Solution of Ammonium Sulfate

Calculate the mass of ammonium sulfate required according to the theoretical output of ferrous sulfate, weigh the corresponding amount and prepare it into saturated solution referring to table 3-4.

Table 3-4 Solubility of ammonium sulfate (g/100g H_2O)

t/℃	10	20	30	40	50
Solubility	70.6	73.0	75.4	78.0	81.0

4. The Preparation of Ferrous Ammonium Sulfate

Evaporate the mixture with an alcohol burner or electric furnace until the crystalline film just appears on the surface of the solution. Cool it to room temperature, that is, ammonium ferrous sulfate crystal precipitates, observe the crystal color. The mother liquor and crystal were separated completely as far as possible by suction filtration with a Buchner funnel, and the crystal was washed with a small amount of ethanol. Take out the crystal, put it between two clean filter papers, and press lightly to absorb the mother liquor; weigh and calculate the theoretical yield and productivity.

5. Quality Test of Fe^{3+} (Limit Analysis)

5.1 Prepare the Fe^{3+} standard solution with a concentration of 0.0100mg/ml

Accurately weigh 0.0216g $(NH_4)_2Fe(SO_4)_2 \cdot 6H_2O$ into a beaker, add a small amount of distilled water to dissolve, then add 6ml of 3mol/L H_2SO_4 solution for acidification, and use distilled water to fix the volume of the solution in a 250ml volumetric flask. The concentration of Fe^{3+} in this solution is 0.0100mg/ml.

5.2 Preparation of standard color scale

Pipette 5.00, 10.00 and 20.00ml of Fe^{3+} standard solution into the colorimetric tube, add 1ml 3mol/L of H_2SO_4 and 1ml of 25% KSCN solution respectively, dilute the solution to 25ml with newly boiled and over cooled distilled water, shake it up, and then obtain three grades of reagent standard solution with

Fe^{3+} content of 0.05 mg (first grade), 0.10mg (second grade) and 0.20mg (third grade).

5.3 Determination of product grade

Weigh 1.0g of ammonium ferrous sulfate crystal, add 15ml of distilled water without oxygen for dissolution, transfer it to 25ml colorimetric tube quantitatively, add 1ml of 3mol/L H_2SO_4 and 1ml of 25% KSCN solution, add distilled water without oxygen to dilute the solution to 25ml, shake it up, compare with standard solution visually, and determine the product grade.

Notes

1. During the preparation of $FeSO_4$, the water bath temperature should not exceed 80℃, so as not to react too violently.

2. During the preparation of $FeSO_4$, keep the pH of the solution no more than 1, so that the reaction between iron filings and sulfuric acid solution can continuously carry out.

3. During testing Fe^{3+} content in products, in order to prevent Fe^{3+} from being oxidized by oxygen dissolved in water, distilled water can be heated to boiling to drive out the oxygen dissolved in water.

4. During the preparation of ferrous ammonium sulfate crystal, the solution must be acidic without stirring, and not be condensed to dry during evaporation and concentration.

Questions

(1) What should be paid attention to when heating the water bath?

(2) How to determine the required amount of ammonium sulfate? How to prepare saturated solution of ammonium sulfate?

(3) Why excess iron is needed in the preparation of ferrous ammonium sulfate?

(4) Why should we keep the strong acidity of the solution when preparing ferrous ammonium sulfate?

实验十三　葡萄糖酸锌的制备

实验目的

1. **掌握**　葡萄糖酸锌的制备方法。
2. **熟悉**　药品质量分析中的滴定分析方法。

实验原理

葡萄糖酸锌，分子式为 $C_{12}H_{22}O_{14}Zn$，为白色晶体或颗粒性粉末，无臭，味微涩。溶于水，极易溶于沸水，不溶于无水乙醇、三氯甲烷或乙醚中。常用的制备方法是葡萄糖酸钙与硫酸锌直接反应制得：

$$[CH_2OH(CHOH)_4COO]_2Ca + ZnSO_4 \rightarrow [CH_2OH(CHOH)_4COO]_2Zn + CaSO_4$$

过滤除去硫酸钙沉淀，溶液经浓缩结晶可得葡萄糖酸锌晶体。

葡萄糖酸锌在作为药物前，要进行多个项目的检测。本实验采用 EDTA 配位滴定法测定葡萄糖酸锌含量，根据 EDTA 与锌的配位反应，由 EDTA 标准溶液消耗的体积计算葡萄糖酸锌的含量。《中国药典》（2020 年版）规定葡萄糖酸锌含量应为 97.0%~102%。

仪器、试剂及其他

1. **仪器**　烧杯（100ml），量筒（50ml），移液管，容量瓶，蒸发皿，水浴锅，减压抽滤装置，普通漏斗，电子分析天平，锥形瓶。
2. **试剂**　葡萄糖酸钙，硫酸锌（$ZnSO_4 \cdot 7H_2O$），95% 乙醇，活性炭，乙二胺四乙酸二钠滴定液，铬黑 T 指示剂。

实验内容

1. **葡萄糖酸锌的制备**　量取 40ml 蒸馏水置 100ml 的烧杯中，加热至 80~90℃，加入 6.7g $ZnSO_4 \cdot 7H_2O$ 使其完全溶解，将烧杯放在 90℃的恒温水浴中，再逐渐加入葡萄糖酸钙 10g，并不断搅拌。在 90℃水浴中保温 20 分钟后趁热抽滤，滤液移至蒸发皿中并在沸水浴上浓缩至黏稠状（体积约为 10ml，如浓缩液有沉淀，需过滤）。滤液冷却至室温，加 95% 乙醇 20ml 并不断搅拌，此时有大量的胶状葡萄糖酸锌析出。充分搅拌后，用倾析法去除乙醇液。再在沉淀上加 95% 乙醇 20ml，充分搅拌后，沉淀慢慢转变成晶体状，抽滤至干，即得粗品（母液回收）。

将粗品加 20ml 蒸馏水，加热至溶解，趁热抽滤，滤液冷却至室温，加 95% 乙醇 20ml 充分搅拌，结晶析出后，抽滤至干，即得精品，50℃烘干。

2. **葡萄糖酸锌的含量测定**　精密称取本品约 0.7g，加蒸馏水 20ml，微热使溶解，定容至 100ml，准确移取 25.00ml 至锥形瓶中，加氨－氯化铵缓冲液 (pH10.0) 10ml 与铬黑 T 指示剂 4 滴，用 EDTA 标准溶液 (0.05mol/L) 滴定至溶液自紫红色转变为纯蓝色，即为终点。平行测定 3 份，记录 EDTA-2Na 标准液消耗的体积，计算葡萄糖酸锌的含量。按照《中国药典》规定每 1ml EDTA-2Na 标准滴定液（0.05mol/L）相当于 22.78mg 的葡萄糖酸锌。

3. 实验记录和结果计算（表 3-5）

表 3-5 葡萄糖酸锌的含量测定

测定次数	1	2	3
葡萄糖酸锌（g）			
$V_{EDTA-2Na}$ (ml)			
$C_{12}H_{22}O_{14}Zn$ 的含量 (%)			
葡萄糖酸锌平均含量 (%)			

注意事项

1. 不要在沸腾的溶液中加入活性炭，因为会有严重的颠簸。一般情况下，活性炭的使用温度为 75~80℃。
2. 在固体物质完全溶解之前，不应加入活性炭脱色。

思考题

1. 用活性炭脱色应如何操作？
2. 氯化锌或碳酸锌与葡萄糖酸钙反应是否可以制备葡萄糖酸锌？
3. 葡萄糖酸锌含量测定结果若不符合规定，可能由哪些原因引起？

Experiment 13 Preparation of Zinc Gluconate

Objectives

1. Master the preparation method of zinc gluconate
2. Understand the titration analysis method in drug quality analysis

Principle

Zinc gluconate, with a molecular formula of $C_{12}H_{22}O_{14}Zn$, white crystal or granular powder, odorless, slightly astringent. It is soluble in water and very soluble in boiling water, but not in anhydrous ethanol, chloroform or ether. The common preparation method is the direct reaction of calcium gluconate and zinc sulfate to obtain:

$$Ca[CH_2OH(CHOH)_4COO]_2 + ZnSO_4 \rightarrow Zn[CH_2OH(CHOH)_4COO]_2 + CaSO_4 \downarrow$$

The precipitate of calcium sulfate can be removed by filtration and the zinc gluconate crystal can be obtained by concentrated and crystalline solution.

Zinc gluconate must be tested for multiple items before it can be used as a drug. In this experiment, the content of zinc gluconate was determined by EDTA coordination titration method. According to the coordination reaction between EDTA-2Na and zinc, the content of zinc gluconate was calculated from the volume consumed by EDTA-2Na standard solution. The *Chinese Pharmacopoeia* (2020 Ⅱ) stipulates that the zinc gluconate content should be 97.0%~102%.

Equipment, Chemicals and Others

1. Equipment

Beaker, measuring cylinder (50ml), pipette (25ml), volumetric flask (100ml), evaporating dish, water bath, decompression filtration device, funnel, electronic analytical balance, conical flask.

2. Reagents

Zinc gluconate, $ZnSO_4 \cdot 7H_2O$, 95% ethanol, activated carbon, EDTA-2Na, chrome black-T indicator.

Experimental Procedures

1. Preparation of Zinc Gluconate

Weigh 6.7g of $ZnSO_4 \cdot 7H_2O$ to make it completely dissolved in 40ml of distilled water at 90℃ in a 100ml beaker. Put the beaker into a constant temperature water bath at 90℃, and then gradually add 10g of calcium gluconate to the $ZnSO_4$ solution with stirring. Maintain the mixture at 90℃ for 20 minutes, and then filter into an evaporating dish while it is hot. Concentrate the filtrate on the boiling water bath until sticky (the volume is about 10ml if the concentrate has precipitate, need to be filtered). When the filtrate is cooled to room temperature, 20ml of 95% ethanol should be added to the filtrate with stirring. A large amount of colloidal zinc gluconate precipitates out. Remove the ethanol solution by decantation after full agitation. Adding 20ml of 95% ethanol to the precipitate, after full stirring, the precipitate is slowly transformed into crystal shape. That is the crude product after filtering and drying the precipitate.

Add the crude product to 20ml of distilled water heating to dissolve. Filter the solution while it is hot. The filtrate was evenly mixed with 20ml of 95% ethanol after cooling to room temperature. Completely precipitated crystals are filtered to dry and dried at 50℃.

2. Content Determination of Zinc Gluconate

Accurately weigh the product about 0.7g in a beaker. Add 20ml of distilled water to dissolve it by micro-heat, and fix the volume to 100ml. Transfer 25.00ml to a conical flask. Add 10ml of ammonia-ammonium chloride buffer (pH 10.0) with 4 drops of chrome black T indicator to the solution. Titrate with EDTA-2Na standard titrating solution (0.05mol/L) until the solution changes from purplish red to pure blue. Repeat the above operation 3 times. Recorded the consumption volume of EDTA-2Na standard solution, and calculate the zinc gluconate content. According to the *Chinese Pharmacopoeia*, every 1ml of EDTA-2Na standard titration solution (0.05mol/L) is equivalent to 22.78mg of zinc gluconate.

3. Data Recording and Processing

Table 3-5 Determination of zinc gluconate

Testing times	1	2	3
Quality of zinc gluconate (g)			
$V_{EDTA-2Na}$ (ml)			
$C_{12}H_{22}O_{14}Zn$ content (%)			
Average content of zinc gluconate			

Attentions

1. Don't add activated carbon to the boiling solution, as there will be serious bumping. Generally, the service temperature of activated carbon is 75~80℃.

2. Decolorization with activated carbon should not be added until the solid matter is completely dissolved.

Questions

(1) How to decolorize with activated carbon?

(2) Whether zinc gluconate can be prepared by reaction of zinc chloride or zinc carbonate with calcium gluconate?

(3) What are the possible reasons if the determination of zinc gluconate does not meet the requirements?

实验十四　三草酸合铁（Ⅲ）酸钾的制备和性质

实验目的

1．了解电导率法测定物质离子类型的原理和方法。
2．学习一种配合物的制备方法。
3．进一步熟练基础化学实验的基本操作。
4．培养综合研究性实验的能力。

实验原理

三草酸合铁（Ⅲ）酸钾 $K_3[Fe(C_2O_4)_3]\cdot 3H_2O$ 为翠绿色单斜晶体，加热至110℃失去全部结晶水，230℃分解，550℃时分解产物为三氧化二铁与碳酸钾。本实验以硫酸亚铁铵为原料，加草酸钾制得草酸亚铁后经氧化制得三草酸合铁（Ⅲ）酸钾。主要反应式为：

$$(NH_4)_2Fe(SO_4)_2\cdot 6H_2O + H_2C_2O_4 = FeC_2O_4\cdot 2H_2O\downarrow + (NH_4)_2SO_4 + H_2SO_4 + 4H_2O$$

$$6FeC_2O_4\cdot 2H_2O + 3H_2O_2 + 6K_2C_2O_4 = 4K_3[Fe(C_2O_4)_3] + 2Fe(OH)_3\downarrow + 12H_2O$$

$$2Fe(OH)_3 + 3H_2C_2O_4 + 3K_2C_2O_4 = K_3[Fe(C_2O_4)_3] + 6H_2O$$

将溶液蒸发浓缩，加入乙醇后冷却，即析出 $K_3[Fe(C_2O_4)_3]\cdot 3H_2O$ 晶体。

配合物的离子类型可用电导率法来确定。电解质电导率 κ 随溶液中离子数目的不同而变化，即随溶液浓度的不同而变化。通常用摩尔电导率 Λm 来衡量电解质溶液的导电能力。摩尔电导率与电导率之间有如下关系：

$$\Lambda m = \frac{\kappa}{c}\times 10^{-3}$$

式中，κ 为溶液电导率，$S\cdot m^{-1}$；c 为电解质溶液浓度，mol/L；Λm 为含有1mol 电解质溶液的导电能力，$S\cdot m^2\cdot mol^{-1}$。

如果测得一系列已知离子总数的配合物溶液的摩尔电导率，并和被测配合物的摩尔电导率相比较，即可求得该配合物的离子总数，进而可确定该配合物的离子类型。表3-6列出了在25℃时，各种离子类型的化合物在稀度（浓度的倒数）为1024时的摩尔电导率（以 Λ_{1024} 表示）的范围。

表3-6　化合物类型与摩尔电导率（25℃）

化合物类型	MA 型	M_2A 型 MA_2 型	M_3A 型 MA_3 型	M_4A 型 MA_4 型
$\Lambda_{1024}(\times 10^{-4})S\cdot m^2\cdot mol^{-1}$	118~131	235~273	408~442	523~553

配合物的组成可通过化学分析方法确定。其中结晶水的含量可用重量法测得，$C_2O_4^{2-}$ 的含量可用 $KMnO_4$ 标准溶液在酸性介质中测定，Fe^{3+} 的含量则可先用过量的锌粉将其还原为 Fe^{2+}，然后用 $KMnO_4$ 标准溶液滴定。反应如下：

$$5C_2O_4^{2-} + 2MnO_4^- + 16H^+ = 10CO_2 + 2Mn^{2+} + 8H_2O$$

$$5Fe^{2+} + MnO_4^- + 8H^+ = 5Fe^{3+} + Mn^{2+} + 4H_2O$$

仪器、试剂及其他

1. 仪器　恒温水浴锅，循环水式多用真空泵，电子天平，酸式滴定管（50ml），容量瓶（50ml），酒精灯，温度计，布氏漏斗，表面皿等。

2. 试剂　$(NH_4)_2Fe(SO_4)_2 \cdot 6H_2O$，$H_2C_2O_4$（饱和溶液），$K_2C_2O_4$（饱和溶液），$H_2O_2$ (6%)，H_2SO_4 (3mol/L)，KSCN (0.1mol/L)，$FeCl_3$ (0.1mol/L)，$K_3[Fe(CN)_6]$ (0.5mol/L)，$CaCl_2$ (0.1mol/L)，$BaCl_2$ (0.1mol/L)，H_2O_2 (3%)，锌粉，$KMnO_4$ (0.02mol/L、标准液)，四苯硼酸钠（饱和溶液）。

实验内容

1. 三草酸合铁（Ⅲ）酸钾的制备　在200ml烧杯中，将5.0g $(NH_4)_2Fe(SO_4)_2 \cdot 6H_2O$ (s)溶于15ml经3mol/L H_2SO_4（3滴）酸化的水中，加热使其溶解。然后，在不断搅拌下加入25ml饱和 $H_2C_2O_4$ 溶液，将其加热沸腾后静置。待黄色沉淀完全沉降后，倾去上层清夜，用热水洗涤沉淀3~4次至溶液呈中性。在上述沉淀中加入10ml饱和 $K_2C_2O_4$ 溶液，水浴恒温维持40℃左右，缓慢滴加10ml 6% H_2O_2 溶液，此时溶液有红褐色的 $Fe(OH)_3$ 沉淀生成。加热溶液煮沸30~40秒，分解剩余的 H_2O_2。接着将溶液置于沸水浴中，不断搅拌下逐滴滴加饱和 $H_2C_2O_4$ 溶液至溶液pH为3~4。此时溶液呈透明的翠绿色。将溶液在水浴中浓缩至20ml左右，趁热过滤。

将滤液放入装有干燥剂的干燥器内并放入暗处使其自然冷却结晶。结晶完全后，抽滤，用无水乙醇淋洗晶体，抽干，低温干燥，称重，计算产率。

2. 配合物的性质

（1）配合物离子类型的测定（电导率法）

① 配制稀度（1/c）为256的产物溶液，用100ml的容量瓶定容。自行计算所需称取产物的量（标准至0.1mg）。

② 用移液管移取稀度为256的产物溶液25ml于100ml的容量瓶中，稀释至刻度线，摇匀，即得稀度为1024的产物溶液。

③ 测定溶液的电导度：将稀度为1024的产物溶液倒入洁净、干燥的小烧杯中，用电导仪测定溶液的电导率κ，计算摩尔电导率 Λ_{1024}，判断三草酸合铁（Ⅲ）酸钾配合物的离子类型。

（2）配合物外界离子的测定　称取1g的产物溶于10ml的蒸馏水中，配成产物的饱和溶液。

① 鉴定 K^+：取少量饱和 $K_2C_2O_4$ 溶液和产物的饱和溶液于两支试管中，分别加入几滴四苯硼酸钠饱和溶液，充分摇匀，观察有无白色沉淀产生。

② 鉴定 Fe^{3+}：取少量 0.1mol/L Fe^{3+} 溶液和产物的饱和溶液于两支试管中，分别加入0.1mol/L KSCN溶液1滴，观察现象有何不同？在盛有产物溶液的试管中加入3mol/L H_2SO_4 溶液2滴，溶液颜色有何变化？解释实验现象。

③ 鉴定 $C_2O_4^{2-}$：取少量饱和 $K_2C_2O_4$ 溶液和产物的饱和溶液于两支试管中，分别加入0.1mol/L $CaCl_2$ 溶液两滴，观察现象有何不同？放置一段时间后又有何变化？解释实验现象。

综合以上实验现象，确定所制的配合物中哪种离子在内界？哪种离子在外界？将实验结果记录在表3-7中。

表 3-7 配合物内外界离子的测定

待检离子	检验方法	现象（产物溶液）
K^+	加四苯硼酸钠饱和溶液	
Fe^{3+}	加 KSCN 溶液	
	加 H_2SO_4 溶液	
$C_2O_4^{2-}$	加 $CaCl_2$ 溶液	
	放置	

3. 配合物的组成分析

（1）结晶水含量的测定

① 洗净两个 φ2.5cm×4.0cm 的称量瓶，放入烘箱中，在110℃下干燥1小时，置于干燥器中冷却至室温，称量。重复上述干燥（0.5小时）→冷却→称量等操作至恒重（两次称量相差不超过0.3mg）。

② 准确称取 0.9~1.0g 晾干的产物两份（称准至 0.1mg），分别放入上述两个已恒重的称量瓶中。半开称量瓶盖，放入烘箱中，在110℃下干燥1小时，置于干燥器中冷却至室温，称量。重复上述干燥（0.5小时）→冷却→称量等操作，直至恒重。根据称量结果，计算每克无水配合物所对应结晶水的质量分数 $\omega(H_2O)$。

$$\omega(H_2O) = \frac{m(H_2O)}{m(无水物)} = \frac{m(产物) - m(无水物)}{m(无水物)}$$

将实验结果记录在表 3-8 中。

表 3-8 结晶水含量的测定

实验序号		Ⅰ	Ⅱ
m(称量瓶)/g			
m(称量瓶+产物)/g			
m(产物)/g			
m(称量瓶+无水物)/g			
m(无水物)/g			
$m(H_2O)$/g			
$\omega(H_2O)$	测定值		
	平均值		

（2）$C_2O_4^{2-}$ 含量的测定　准确称取 0.15~0.20g 无水产物两份（标准至 0.1mg），置于锥形瓶中，加入 50ml 蒸馏水和 10ml 3mol/L H_2SO_4 溶液，微热溶解。

在锥形瓶中先用滴定管滴加约 10ml 0.02mol/L $KMnO_4$ 标准溶液（是否需要计入读数?），加热至 70~80℃（冒泡较多水蒸气，锥形瓶壁上有回流），趁热在继续用 $KMnO_4$ 标准溶液滴定，开始滴定时要慢并摇动均匀，待紫红色褪去后再加入第二滴，整个滴定过程保持温度不低于60℃，滴定溶液呈粉红色并在30秒内不褪色。记录 $KMnO_4$ 标准溶液的用量 V_1。计算每克无水配合物所含 $C_2O_4^{2-}$ 的质量分数 $\omega(C_2O_4^{2-})$。

滴定反应式　　　$5C_2O_4^{2-} + 2MnO_4^- + 16H^+ = 10CO_2 + 2Mn^{2+} + 8H_2O$

$$\omega(C_2O_4^{2-}) = \frac{\frac{5}{2}c(KMnO_4)V_1(KMnO_4)M(C_2O_4^{2-})}{m(无水物) \times 1000}$$

保留滴定后的溶液，用作 Fe^{3+} 的测定。

（3）Fe^{3+} 含量的测定　将上述滴定后的溶液加热近沸，加入半药匙锌粉，直至溶液的黄色消失（如何解释）。用短颈漏斗趁热将溶液过滤于另一锥形瓶中，用 5ml 蒸馏水通过漏斗洗涤残渣一次，洗涤液与滤液合并收集于同一锥形瓶中。用 $KMnO_4$ 标准溶液滴定至溶液呈粉红色并在 30 秒内不褪色，记录 $KMnO_4$ 标准溶液的用量 V_2。计算每克无水配合物所含 Fe^{3+} 的质量分数 $\omega(Fe^{3+})$。

滴定反应式　　　　　　$5Fe^{2+} + MnO_4^- + 8H^+ = 5Fe^{3+} + Mn^{2+} + 4H_2O$

$$\omega(Fe^{3+}) = \frac{5c(KMnO_4)V_2(KMnO_4)M(Fe^{3+})}{m(无水物) \times 1000}$$

用另一份样品重复上述测定，将实验结果记录在表 3-9 中。

表 3-9　配合物组成的分析

实验序号		Ⅰ	Ⅱ
m(无水物)/g			
$c(KMnO_4)$/mol/L			
$V_1(KMnO_4)$/ml			
$\omega(C_2O_4^{2-})$	测定值		
	平均值		
$V_2(KMnO_4)$/ml			
$\omega(Fe^{3+})$	测定值		
	平均值		

4. 产物化学式的确定　由实验所测得的 $\omega(H_2O)$、$\omega(C_2O_4^{2-})$ 和 $\omega(Fe^{3+})$ 值，可计算每克无水产物所含 K^+ 的质量分数 $\omega(K^+)$。据此可计算配合物中各组分的物质的量的比：

$$n(K^+) : n(Fe^{3+}) : n(C_2O_4^{2-}) : n(H_2O)$$

再结合配合物的离子类型和配合物内、外界，即可确定配合物的化学式。

5. 数据记录与结果处理

室温_____℃

硫酸亚铁铵的质量_____g

三草酸合铁（Ⅲ）酸钾的理论产量_____g

三草酸合铁（Ⅲ）酸钾的实际产量_____g

产率_____

电导率电极的常数值_____cm^{-1}

配合物溶液电导率 $\kappa =$ _____$S \cdot m^2 \cdot mol^{-1}$

配合物的离子类型_____

配合物的外界离子_____，配合物的内界离子_____

每克无水配合物所含 K^+ 的质量分数 $\omega(K^+)=$ _____

配合物中各组分的物质的量之比：_____

配合物的化学式_____

注意事项

1. 第一次加入饱和 $H_2C_2O_4$ 时要注意搅拌，防止暴沸。

2. 加 3% H_2O_2 要少量多次，且慢搅拌。加热虽能加快非均相反应的速率，但也将促使 H_2O_2 分解，故温度不宜太高。

3. $KMnO_4$ 滴定 $C_2O_4^{2-}$ 时溶液的温度不能超过 90℃，否则，部分 $H_2C_2O_4$ 会分解。

$H_2C_2O_4 \rightarrow CO_2\uparrow + CO\uparrow + H_2O$，但滴定时的温度不能低于 60℃，即使是在 70~80℃的温度下，滴定反应的速率仍然很慢，$KMnO_4$ 溶液必须逐滴加入（第一滴加入后，要摇匀溶液，当紫色褪去后再加第二滴）。否则，加入的 $KMnO_4$ 溶液来不及与 $C_2O_4^{2-}$ 反应，在热的酸性溶液中会发生分解：

$$4MnO_4^- + 12H^+ = 4Mn^{2+} + 5O_2\uparrow + 6H_2O$$

4. 锌粉除了与 Fe^{3+} 反应外，也与溶液中的 H^+ 反应，所以加入锌粉需过量，同时溶液必须保持足够的酸度，以免 Fe^{3+}、Fe^{2+}、Zn^{2+} 等水解而析出。

5. 接近终点时，紫红色褪去很慢，应减慢滴定速度，同时充分振荡，以防超过终点。

思考题

1. 影响三草酸合铁（Ⅲ）酸钾产量的主要因素有哪些？
2. 应根据哪种试剂的量计算产率？
3. 如何提高产率？能否用蒸干溶液的办法来提高产率？
4. 在实验中加入乙醇的作用是什么？
5. 用 $KMnO_4$ 标准溶液滴定配合物中的 $C_2O_4^{2-}$ 时，为什么要先加入约 10ml 的 $KMnO_4$ 标准溶液，并在加热开始反应后再滴定？先加入的 $KMnO_4$ 溶液是否需要准确读数？

Experiment 14   Preparation and Properties of Potassium Trioxalatoferrate (Ⅲ) Trihydrate

Objectives

1 Understand the principle and method of determining the ion type of a substance by conductivity method.

2. Learn how to prepare a complex.

3. Familiar with some basic operations of basic chemistry experiments.

4. Cultivate the ability of comprehensive research experiments.

Principles

Potassium Trioxalatoferrate (Ⅲ) Trihydrate $\{K_3[Fe(C_2O_4)_3] \cdot 3H_2O\}$ is an emerald green monoclinic crystal, heated to 110℃ to lose all crystal water, decomposed at 230℃, and the decomposition product at 550℃ is ferric oxide and potassium carbonate.

In this experiment, ferrous ammonium sulfate was used as a raw material, potassium oxalate was added to obtain ferrous oxalate, and then iron (Ⅲ) potassium trioxalate was obtained by oxidation. The main reactions formulas are:

$$(NH_4)_2Fe(SO_4)_2 \cdot 6H_2O + H_2C_2O_4 = FeC_2O_4 \cdot 2H_2O \downarrow + (NH_4)_2SO_4 + H_2SO_4 + 4H_2O$$

$$6FeC_2O_4 \cdot 2H_2O + 3H_2O_2 + 6K_2C_2O_4 = 4K_3[Fe(C_2O_4)_3] + 2Fe(OH)_3 \downarrow + 12H_2O$$

$$2Fe(OH)_3 + 3H_2C_2O_4 + 3K_2C_2O_4 = K_3[Fe(C_2O_4)_3] + 6H_2O$$

The solution was concentrated by evaporation, cooled after adding ethanol, and $K_3[Fe(C_2O_4)_3] \cdot 3H_2O$ crystals were precipitated.

The ion type of the complex can be determined by the conductivity method. The conductivity (κ) of the electrolyte varies with the number of ions in the solution, that is, with the concentration of the solution. The electrical conductivity of an electrolyte solution is usually measured by the molar conductivity (Λm). The relationship between molar conductivity and conductivity is as follows:

$$\Lambda m = \frac{\kappa}{c} \times 10^{-3}$$

In the formula, κ is the conductivity of the solution, $S \cdot m^{-1}$; c is the concentration of the electrolyte solution, mol/L; Λm is the conductivity of the electrolyte solution containing 1mol, $S \cdot m^2 \cdot mol^{-1}$.

If the molar conductivity of a series of substances with known ion numbers is measured and compared with the molar conductivity of the tested complex, the total number of ions of the complex can be obtained, and the ion type of the complex can be determined. Table 3-6 lists the range of molar conductivity (represented by Λ_{1024}) of various ion types of compounds at a diluteness (reciprocal of concentration) of 1024 at 25℃.

Table 3-6 Compound types and molar conductivity (25℃)

Compound type	MA type	M$_2$A type MA$_2$ type	M$_3$A type MA$_3$ type	M$_4$A type MA$_4$ type
$\Lambda_{1024}(\times 10^{-4})$S·m^2·mol^{-1}	118~131	235~273	408~442	523~553

The composition of the complex can be determined by chemical analysis. The content of crystal water can be measured by gravimetric method. The content of $C_2O_4^{2-}$ can be measured in $KMnO_4$ standard solution in an acidic medium. The content of Fe^{3+} can be reduced to Fe^{2+} with excess zinc powder first, and then titrated with KMnO4 standard solution. The reactions are as follows:

$$5C_2O_4^{2-}+2MnO_4^-+16H^+ = 10CO_2+2Mn^{2+}+8H_2O$$
$$5Fe^{2+}+MnO_4^-+8H^+ = 5Fe^{3+}+Mn^{2+}+4H_2O$$

 Equipment, Chemicals and Others

1. Equipment

Constant temperature water bath pot, circulating water multi-purpose vacuum pump, electronic balance, acid burette (50ml), volumetric flask (50ml), alcohol burner, thermometer, buchner funnel, watch glass, etc.

2. Chemicals

$(NH_4)_2Fe(SO_4)_2 \cdot 6H_2O$ (homemade), $H_2C_2O_4$ (saturated solution), $K_2C_2O_4$ (saturated solution), H_2O_2 (6%), H_2SO_4 (3mol/L), KSCN (0.1mol/L), $FeCl_3$ (0.1mol/L), $K_3[Fe(CN)_6]$ (0.5mol/L), $CaCl_2$ (0.1mol/L), $BaCl_2$ (0.1mol/L), zinc powder, $KMnO_4$ (0.02mol/L, standard solution), sodium tetraphenylboron (saturated solution).

 Experimental Procedures

1. Preparation of Potassium Trioxalatoferrate (Ⅲ) Trihydrate

In a 200ml beaker, 5.0g $(NH_4)_2Fe(SO_4)_2 \cdot 6H_2O$ (s) was dissolved in 15ml of water acidified with 3mol/L H_2SO_4 (3 drops), and heated to dissolve. Then, add 25ml of saturated $H_2C_2O_4$ solution under constant stirring, heat it to boil, and let it stand. After the yellow precipitate has completely settled, the upper layer is decanted, and the precipitate is washed with hot water 3~4 times until the solution becomes neutral. Add 10ml of saturated $K_2C_2O_4$ solution to the above precipitate, maintain the water bath at a constant temperature of about 40℃, and slowly add 10ml of 6% H_2O_2 solution dropwise. At this time, a red-brown $Fe(OH)_3$ precipitate was formed in the solution. Boil the solution for 30~40 seconds to decompose the remaining H_2O_2. Then the solution is placed in a boiling water bath, and the saturated $H_2C_2O_4$ solution is added dropwise with constant stirring until the solution pH is 3~4. The solution was now transparent emerald green. The solution was concentrated to about 20ml in a water bath and filtered while hot. The filtrate was placed in a desiccator with a desiccant and placed in a dark place to allow natural cooling to crystallize. After the crystallization is complete, the crystals are suction filtered, the crystals are rinsed with absolute ethanol, dried with suction, dried at low temperature, weighed and the yield is calculated.

2. Properties of the Complex

2.1 Determination of complex ion type (conductivity method)

(1) Prepare a product solution with a dilution (1/c) of 256 and make up to a 100ml volumetric flask. Calculate the amount of product you need to weigh (standard to 0.1 mg).

(2) Using a pipette, transfer 25ml of a product solution with a dilution of 256 into a 100ml volumetric flask, dilute to a tick mark, and shake to obtain a product solution with a dilution of 1024.

(3) Determining the electrical conductivity of a solution. The product solution with a dilution of 1024 was poured into a clean, dry small beaker, the conductivity (κ) of the solution was measured with a conductivity meter, the molar conductivity (Λ) was calculated to be 1024, and the ion type of the potassium trioxalate (Ⅲ) acid complex was determined.

2.2 Determination of external ions of the complex

Weigh 1g of the product and dissolve it in 10ml of distilled water to prepare a saturated solution of the product.

(1) Identification K^+ Take a small amount of a saturated $K_2C_2O_4$ solution and a saturated solution of the product into two test tubes, add a few drops of saturated sodium tetraphenylborate solution, shake them thoroughly, and observe whether white precipitation occurs.

(2) Identification Fe^{3+} Take a small amount of the 0.1mol/L Fe^{3+} solution and a saturated solution of the product into two test tubes, and add one drop of the 0.1mol/L KSCN solution. Observe what is the difference? Add two drops of 3mol/L H_2SO_4 solution to the test tube containing the product solution. How does the color of the solution change? Explain the experimental phenomena.

(3) Identification of $C_2O_4^{2-}$ Take a small amount of saturated $K_2C_2O_4$ solution and the saturated solution of the product in two test tubes, and add two drops of 0.1mol/L $CaCl_2$ solution, and observe what is the difference? What happens after a period of time? Explain the experimental phenomena.

Based on the above experimental phenomena, what kind of ions in the prepared complex is determined to be in the internal boundary? Which ion is outside? The experimental results are recorded in Table 3-7.

Table 3-7 Determination of external and internal ions in complexes

Test ions	Testing method	Phenomenon (product solution)
K^+	Add saturated sodium tetraphenylborate solution	
Fe^{3+}	Add KSCN solution	
	Add H_2SO_4 solution	
$C_2O_4^{2-}$	Add $CaCl_2$ solution	
	Stand still	

3. Composition analysis of complex

3.1 Determination of crystal water content

(1) Wash two weighing bottles with a diameter of φ 2.5cm ×4.0cm, put them in an oven, and dry them at 110℃ for 1h. Then put them in a desiccator to cool to room temperature and weigh. Repeat the above drying (0.5h) → cooling → weighing to constant weight (the difference between the two weighing shall not exceed 0.3mg).

(2) Weigh accurately 0.9~1.0g of the dried product in two parts (accurate to 0.1 mg), and place them

into the above two constant-weight weighing bottles. Open the bottle cap halfway, put it in an oven, and dry it at 110℃ for 1 hour, put it in a desiccator to cool to room temperature, and weigh. Repeat the above drying (0.5h) → cooling → weighing and other operations until constant weight. According to the weighing result, calculate the mass fraction $\omega(H_2O)$ of the crystal water corresponding to each gram of anhydrous complex.

$$\omega(H_2O) = \frac{m(H_2O)}{m(\text{anhydrous})} = \frac{m(\text{product}) - m(\text{anhydrous})}{m(\text{anhydrous})}$$

The experimental results are recorded in Table 3-8.

Table 3-8 Determination of crystal water content

Experiment number		I	II
m(Weighing bottle)/g			
m(Weighing bottle+product)/g			
m(product)/g			
m(Weighing bottle+anhydrous)/g			
m(anhydrous)/g			
$m(H_2O)$/g			
$\omega(H_2O)$	measured value		
	average value		

3.2 Determination of $C_2O_4^{2-}$ Contend

Weigh accurately 0.15~0.20g of two anhydrous products (standard to 0.1mg), place them in a conical flask, add 50ml of distilled water and 10ml of 3mol/L H_2SO_4 solution, and dissolve slightly.

In a conical flask, first add about 10ml of 0.02mol/L $KMnO_4$ standard solution with a burette (do you need to count the reading?), And then heat to 70~80℃ (more water vapor bubbles, there is reflux), continue to titrate with $KMnO_4$ standard solution while it is still hot, start the titration slowly and shake evenly, add a second drop after the purple red fades, maintain the temperature during the entire titration process not lower than 60℃, the titration solution is pink and does not fade in 30 seconds. Record the amount of $KMnO_4$ standard solution V_1. Calculate the mass fraction $\omega(C_2O_4^{2-})$ of $C_2O_4^{2-}$ per gram of anhydrous complex.

Titration reaction $5C_2O_4^{2-} + 2MnO_4^- + 16H^+ = 10CO_2 + 2Mn^{2+} + 8H_2O$

$$\omega(C_2O_4^{2-}) = \frac{\frac{5}{2}c(KMnO_4)V_1(KMnO_4)M(C_2O_4^{2-})}{m(\text{anhydrous}) \times 1000}$$

The titrated solution was retained for the determination of Fe^{3+}.

3.3 Determination of Fe^{3+} content

Heat the titrated solution to near-boiling and add half-key zinc powder until the yellow color of the solution disappears (how to explain?). The solution was filtered in a conical flask with a short-necked funnel while it was still hot, and the residue was washed through the funnel once with 5ml of distilled water. The washing solution and the filtrate were collected in the one conical flask. Titrate with $KMnO_4$

standard solution until the solution is pink and does not fade within 30 seconds. Record the amount of $KMnO_4$ standard solution V_2. Calculate the mass fraction ω (Fe^{3+}) of Fe^{3+} per gram of anhydrous complex.

Titration reaction $\quad\quad\quad 5Fe^{2+}+MnO_4^-+8H^+ = 5Fe^{3+}+Mn^{2+}+4H_2O$

$$\omega(Fe^{3+}) = \frac{5c(KMnO_4)V_2(KMnO_4)M(Fe^{3+})}{m(\text{anhydrous})\times 1000}$$

The above measurement was repeated with another sample, and the experimental results are recorded in Table 3-9.

Table 3-9 Composition analysis of complex

Experiment number		I	II
m(anhydrous)/g			
$c(KMnO_4)$/mol/L			
$V_1(KMnO_4)$/ml			
$\omega(C_2O_4^{2-})$	measured value		
	average value		
$V_2(KMnO_4)$/ml			
$\omega(Fe^{3+})$	measured value		
	average value		

4. Determination of the chemical formula of the product

From the measured values of ω (H_2O), ω ($C_2O_4^{2-}$) and ω (Fe^{3+}), the mass fraction ω (K^+) of K^+ contained in each gram of anhydrous product can be calculated. From this, the ratio of the amount of each component in the complex can be calculated:

$$n(K^+):n(Fe^{3+}):n(C_2O_4^{2-}):n(H_2O)$$

Combining the ionic type of the complex with the inside and outside of the complex, the chemical formula of the complex can be determined.

5. Data logging and results processing

Room temperature _____ ℃

Mass of ferrous ammonium sulfate _____ g

Theoretical yield of potassium trioxalate (III) _____ g

Actual yield of potassium trioxalate (III) _____ g

Yield _____

Constant value of conductivity electrode _____ cm^{-1}

Conductivity of complex solution κ = _____ $S \cdot m^2 \cdot mol^{-1}$

Complex ion type _____

External ion of complex _____, Internal ion of complex _____

The mass fraction of K^+ per gram of anhydrous complex ω (K^+) = _____

The ratio of the amount of each component in the complex: _____

Chemical formula of the complex _____

 Notes

1. When adding saturated $H_2C_2O_4$ for the first time, pay attention to stirring to prevent bumping.

2. Add 3% H_2O_2 in small amounts and many times and stir slowly. Although heating can accelerate the reaction rate of the heterogeneous phase, it will also promote the decomposition of H_2O_2, so the temperature should not be too high.

3. When $KMnO_4$ titrates $C_2O_4^{2-}$, the temperature of the solution cannot exceed 90℃, otherwise, some $H_2C_2O_4$ will decompose; $H_2C_2O_4 \rightarrow CO_2 \uparrow + CO \uparrow + H_2O$, but the temperature during the titration must not be lower than 60℃, even at 70~80℃, the rate of the titration reaction is still very slow, and the $KMnO_4$ solution must be added dropwise (After adding the first drop, shake the solution well and add a second drop when the purple color has faded). Otherwise, the $KMnO_4$ solution added is too late to react with $C_2O_4^{2-}$, and it will decompose in the hot acidic solution:

$$4MnO_4^- + 12H^+ \rightarrow 4Mn^{2+} + 5O_2 \uparrow + 6H_2O$$

4. In addition to reacting with Fe^{3+}, zinc powder also reacts with H^+ in the solution, so the zinc powder needs to be added in excess, and the solution must maintain sufficient acidity to prevent the hydrolysis of Fe^{3+}, Fe^{2+}, Zn^{2+}, etc.

5. When approaching the end point, the fuchsia fades slowly, and the titration speed should be slowed down while shaking sufficiently to prevent the end point from being exceeded.

Questions

(1) What are the main factors that affect the production of potassium trioxalat oferrate (Ⅲ) trihydrate?

(2) Which reagent should be used to calculate the yield?

(3) How to increase productivity? Can the solution be evaporated to increase the yield?

(4) What is the effect of adding ethanol to the experiment?

(5) When using $KMnO_4$ standard solution to titrate $C_2O_4^{2-}$ in the complex, why should I add about 10ml $KMnO_4$ standard solution first, and then titrate after heating to start the reaction? Do I need to read the $KMnO_4$ solution accurately first?

实验十五　无机阴、阳离子的鉴定和矿物药的鉴别

实验目的

1. **掌握**　常见无机阴、阳离子的特征反应和鉴定反应的操作。
2. **熟悉**　芒硝、硝石、白矾、大青盐、硼砂等矿物药的化学鉴定方法。
3. **了解**　未知物定性分析的试验流程。
4. **明确**　化学反应和鉴定反应的关系。

实验原理

利用加入试剂，使其与溶液中的某种无机离子发生特征化学反应，鉴定该离子存在与否的试验称为离子的鉴别。鉴别反应须具有下述特征之一：①沉淀的生成或沉淀的溶解；②溶液或沉淀的颜色变化；③特殊气体的生成并逸出；④产生其他特殊现象等。结晶反应、焰色反应和气室反应常被用作鉴定反应。有机试剂的应用往往能提高离子鉴定反应的特效性和敏感性。

仪器、试剂及其他

1. **仪器**　试管，铂丝，玻璃棒，离心机，酒精灯，试管夹，点滴板，表面皿。
2. **试剂**

酸：浓 HCl，1mol/L HCl，浓 H_2SO_4，2mol/L H_2SO_4，浓 HNO_3。

碱：2mol/L NaOH，2mol/L 氨试液。

盐：K^+、Na^+、Ag^+、Mg^{2+}、Fe^{2+}、Ba^{2+}、Al^{3+}、Sb^{3+}、NO_3^- 离子溶液，硼砂晶体，NH_4^+、As^{3+}、SO_3^{2-} 盐固体，3mol/L NH_4Ac，饱和 $FeSO_4$，1mol/L Na_2S，1mol/L $Hg(NO_3)_2$，1mol/L $AgNO_3$，$NaNO_2$，1mol/L Hg_2Cl_2。

3. **其他**　H_2S 气体，锌粉，四苯硼酸钠，钴亚硝酸钠，醋酸铀酰锌，邻菲罗啉乙醇溶液，镁试剂，铝试剂，玫瑰红酸钠，奈斯勒试剂，甲醇。

实验内容

1. 常见无机阴、阳离子的特征反应

（1）试管试法　将试液放入试管中，滴加试剂，使其产生沉淀或颜色变化的方法叫试管试法。用试管试法进行离子特征反应时，每加入一滴试剂都要充分摇动，直到现象产生。

① Na^+ 盐：取 Na^+ 盐中性溶液 10 滴于小试管中，加醋酸铀酰锌试剂 3~4 滴，不断搅拌并摩擦试管壁，即有黄色沉淀生成，写出离子反应式。

参照上述方法可以鉴别含有 Na^+ 的矿物药，例如，芒硝（主要成分为 $Na_2SO_4 \cdot 10H_2O$），大青盐（主要成分为 NaCl）等。

② K^+ 盐：取 K^+ 盐中性溶液 10 滴于小试管中，加钴亚硝酸钠试剂 3 滴，摇匀后有黄棕色沉淀生成，写出离子反应式。或在另一个试管中加入 K^+ 盐中性溶液 10 滴，加入四苯硼酸钠 3~4 滴，摇匀后观察沉淀的生成，写出离子反应式。

参照上述方法可以鉴别含有 K^+ 的矿物药,例如,硝石(主要成分为 KNO_3),白矾[主要成分为 $KAl(SO_4)_2 \cdot 12H_2O$]等。

③ 硝酸盐:取硝酸盐溶液10滴于小试管中,加入饱和 $FeSO_4$ 溶液 8~12 滴,然后沿着管壁小心加入10滴浓 H_2SO_4 使其成两液层,不要搅动,稍待片刻,观察两液层交界处的颜色,记录颜色环的颜色。

参照上述方法可以鉴别含有 NO_3^- 的矿物药,例如,硝石(主要成分为 KNO_3)等。

④ 硼酸盐:取硼砂晶体约0.5g于小试管中,加蒸馏水3ml,加热溶解,稍冷却后,加浓 H_2SO_4 少许,观察有何变化,加甲醇 5~10ml 后点火燃烧,观察火焰边缘带的颜色。

(2)点滴板反应试法　将少量试液、试剂均滴在点滴板上使其反应的方法叫点滴板反应试法。该试法试液、试剂用量较少。

① Fe^{2+} 盐:取 Fe^{2+} 溶液少许,加1%邻菲罗啉乙醇溶液数滴,观察颜色的变化。反应式如下:

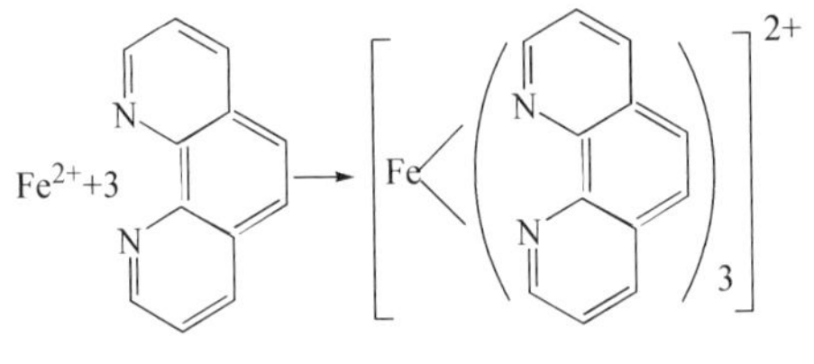

邻啡罗啉　　　橘红色螯合物

② Mg^{2+} 盐:取 Mg^{2+} 溶液少许,加 NaOH 溶液和镁试剂各数滴,观察沉淀的颜色。镁试剂化学名为对硝基苯偶氮对苯二酚,结构式为:

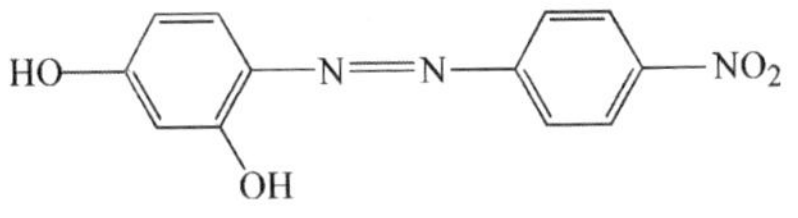

③ Al^{3+} 盐:取 Al^{3+} 盐溶液少许,加数滴 3mol/L NH_4Ac 溶液和 1~2 滴铝试剂,观察沉淀的生成。化学反应为:

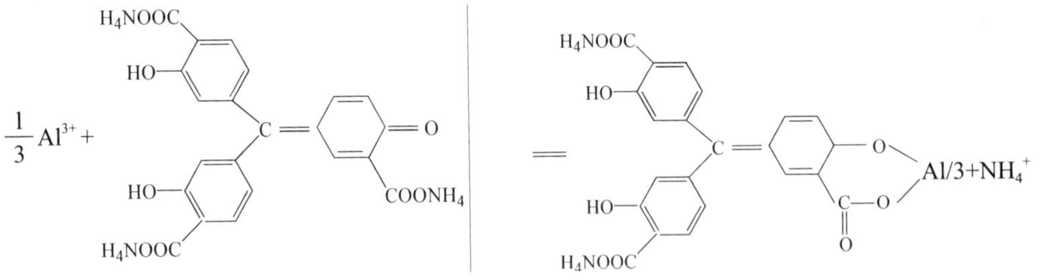

参照上述方法可以鉴别含有 Al^{3+} 的矿物药,例如,白矾[主要成分为 $KAl(SO_4)_2 \cdot 12H_2O$]等。

(3)离心管试法　将试液、试剂放在离心管中进行沉淀反应,并使沉淀物在离心机上离心析出的方法叫离心管试法。开动离心机时,要注意离心机的平稳,离心管放在离心机的管套中要做到对称,必要时用空离心管加水做对称。离心后沉淀物受离心作用沉积在离心管的尖端,可用毛细管将离心液吸出,留下沉淀物在离心管中。

① Ag^+ 盐:取 Ag^+ 溶液少许,加稀盐酸数滴酸化,观察白色凝乳状沉淀的生成,离心,弃去上清液,在沉淀上面滴加氨试液,观察沉淀的溶解。写出离子反应式。

② Sb^{3+} 盐:取 Sb^{3+} 溶液少许,加稀盐酸数滴酸化,通硫化氢气体,观察沉淀的生成,离心,弃去上清液,在沉淀上面滴加硫化钠溶液,观察沉淀的溶解。写出离子反应式。

（4）纸上滴定试法　将试液、试剂滴在滤纸上进行沉淀，由于滤纸的毛细管作用，除沉淀外，其他离子会均匀扩散至沉淀区域以外，沉淀观察比较明显，这种方法叫纸上滴定试法。

Ba^{2+}盐：取Ba^{2+}盐溶液少许滴于滤纸上，加玫瑰红酸钠，观察沉淀的生成，用稀盐酸处理后观察滤纸颜色的变化。

（5）气室试法　气室是由两个小表面皿合在一起构成的，上面一个可稍小并擦干，将试纸润湿后贴在上表面皿的凹面中央，然后盖在下面表面皿上，必要时可放在小烧杯上用蒸气浴加热，待反应发生后观察试纸颜色的变化。这种鉴定离子是否存在的方法叫气室试法。

① NH_4^+盐：取NH_4^+盐固体少许，加过量NaOH试液，加热，分解产生的气体遇浸有奈斯勒试剂的滤纸会产生色斑，观察色斑的生成并写出离子反应式。

② As^{3+}盐：在表面皿上放As^{3+}固体少许，加无砷锌粉少量，加稀硫酸，在气室上表面皿上粘一有硝酸银溶液的滤纸，数分钟后，观察滤纸上色斑的生成。

③ SO_3^{2-}盐：在表面皿上放SO_3^{2-}盐固体少许，加盐酸产生气体，该气体可使由硝酸亚汞溶液润湿的滤纸产生色斑，观察色斑的产生。

（6）焰色试法　将铂丝（或镍铬丝）做成环状，取数滴浓盐酸于点滴板上，将金属环插入盐酸中浸湿，在灯焰上灼烧，如此反复数次直至火焰不染色，表明金属丝已处理洁净。用洁净的金属丝蘸取试液在氧化焰中灼烧，通过火焰特征颜色来鉴别无机离子的方法叫焰色试法。

① 钠盐：在火焰中呈亮黄色。
② 钾盐：在火焰中呈紫色（当混有钠盐时，可透过蓝色钴玻璃观察）。
③ 钙盐：在火焰中呈砖红色。
④ 钡盐：在火焰中呈黄绿色。

记录每种盐在火焰中呈现的焰色，每次实验前用上述方法洁净金属丝。千万不可将铂丝放于还原焰中灼烧，否则生成碳化铂，铂丝会发生脆断。

2. 未知物的鉴别　领取未知溶液一份，其中可能含有的离子是：Na^+、K^+、NH_4^+、Mg^{2+}、Ca^{2+}、Ba^{2+}、Cl^-、Br^-、I^-，参照以上实验，自己拟定分析步骤，确定未知溶液中含有哪些离子。

3. 矿物药的鉴别　领取未知矿物药2份，其中可能含有的离子是：Na^+、K^+、Al^{3+}、Cl^-、NO_3^-、SO_4^{2-}，请设计一个实验方案来鉴别未知矿物药。

思考题

1. 哪些操作技术被用来进行常见无机离子的鉴定反应？
2. 进行离子"一般鉴定反应"应具有什么前提？
3. 若未知液中有Br^-无Cl^-，而在处理卤化银沉淀的2mol/L$NH_3·H_2O$中加HNO_3检出Cl^-时，溶液却变浑浊，试解释这种现象？

Experiment 15 Identification of Inorganic Anions, Cations, and Mineral Drugs

Objectives

1. Understand the characteristic reactions of common inorganic anions and cations, and learn about the operation of identification reactions.
2. Familiar with the chemical identification methods of mirabilite, niter, alum, halite, borax and other mineral drugs.
3. Learn the experimental procedure for qualitative analysis of unknown chemical species.
4. Identify the relationship between chemical reactions and identification reactions.

Principles

The test of adding a reagent to cause a characteristic chemical reaction with an inorganic ion in solution to help identify its existence is called the identification of the ion. The identification reaction must have one of the following characteristics: ① formation or dissolution of precipitates; ② changes in color of the solution or precipitates; ③ generation and evaporation of special gases; ④ production of other special phenomena. Crystallization reaction, flame test, and gas chamber reaction are often used as identification reactions. The application of organic reagents can often improve the specificity and sensitivity of ion identification reactions.

Equipment, Chemicals and Others

1. Equipment

Test tubes, platinum wires, glass rods, centrifuge, alcohol lamp, test tube clamps, drip plate, watch glass.

2. Chemicals

Acids: concentrated hydrochloric acid (concentrated HCl), 1mol/L HCl, concentrated H_2SO_4, 2mol/L H_2SO_4, concentrated HNO_3.

Bases: 2mol/L NaOH, 2mol/L $NH_3 \cdot H_2O$.

Salts: K^+, Na^+, Ag^+, Mg^{2+}, Fe^{2+}, Ba^{2+}, Al^{3+}, Sb^{3+}, NO_3^- solution, Borax crystal, NH_4^+, As^{3+}, SO_3^{2-} salt solid, 3mol/L NH_4Ac, saturated $FeSO_4$, 1mol/L Na_2S, 1mol/L $Hg(NO_3)_2$, 1mol/L $AgNO_3$, $NaNO_2$, 1mol/L Hg_2Cl_2.

3. Others

H_2S gas, zinc powder, sodium tetraphenylborate, sodium hexanitritocobaltate (Ⅲ), uranyl zinc acetate, phenanthroline ethanol solution, magneson, aluminon, rhodizonic acid sodium salt, Nessler's reagent, methanol.

Experimental Procedures

1. The Characteristic Reaction of Common Inorganic Anions and Cations

1.1 Test Tube Method

The method of putting the test solution in a test tube and adding the reagent dropwise to cause precipitation or color change is called the test tube method. When using the test tube method for the ion characteristic reaction, the solution in the tube should be shaken well after adding each drop of reagent until the phenomenon occurs.

(1) Na^+ Salt: Take 10 drops of Na^+ salt neutral solution in a small test tube, add 3~4 drops of zinc uranyl acetate reagent, and constantly stir and rub the test tube wall, a yellow precipitate will be generated. Please write the ion reaction formula.

Referring to the above methods, mineral drugs containing Na^+ can be identified, for example, mirabilite (the main component is $Na_2SO_4 \cdot 10H_2O$), halite (the main component is $NaCl$), etc.

(2) K^+ Salt: Take 10 drops of K^+ salt neutral solution in a small test tube, add 3 drops of sodium hexanitritocobaltate (Ⅲ) reagent, and shake the mixture to form a yellow-brown precipitate, and write the ion reaction formula. Or take 10 drops of K^+ salt neutral solution to another test tube, add 3~4 drops of sodium tetraphenylborate, shake the test tube and observe the formation of precipitates, and write the ion reaction formula.

Referring to the above methods, mineral drugs containing K^+ can be identified, for example, niter (the main component is KNO_3), alum [the main component is $KAl(SO_4)_2 \cdot 12H_2O$], etc.

(3) Nitrate: Add 10 drops of nitrate solution into a small test tube, add 8~12 drops of saturated $FeSO_4$ solution, and then carefully add 10 drops of concentrated H_2SO_4 along the wall of the tube to form two liquid layers. Do not stir, wait a moment, observe the color at the junction of the two liquid layers, and record the color of the color ring.

Referring to the above methods, mineral drugs containing NO_3^- can be identified, for example, niter (the main component is KNO_3), etc.

(4) Borate: Take about 0.5g of borax crystals in a small test tube, add 3ml of distilled water, heat to dissolve. After cool slightly, add a little concentrated H_2SO_4 into the test tube, observe the changes. Add 5~10ml of methanol, ignite and burn, and observe the color of the flame edge band.

1.2 Drip Plate Reaction Method

The method of dropping a small amount of test solution and reagent on the drip plate to make the reaction is called the drip plate reaction test method, which uses less test solution and reagent.

(1) Fe^{2+} Salt: Take a small amount of Fe^{2+} solution, add a few drops of 1% phenanthroline ethanol solution, and observe the color change. The reaction is as follows:

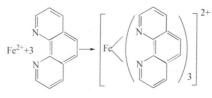

Phenanthroline   Orange Chelate

(2) Mg^{2+} Salt: Take a small amount of Mg^{2+} solution, add several drops of NaOH solution and magneson reagent, observe the color of the precipitate. The chemical name of magneson is 2,

4-Dihydroxy-4'-nitroazobenzene and its structural formula is:

$$\text{HO}\!-\!\!\!\underset{\underset{\text{OH}}{|}}{\bigcirc}\!\!\!-\!N\!=\!N\!-\!\bigcirc\!-\!NO_2$$

(3) Al^{3+} Salt: Take a small amount of Al^{3+} solution, add 3mol/L NH_4Ac solution and 1~2 drops of aluminon, and observe the formation of precipitation. The chemical reaction is:

[Structural equation showing aluminon reacting with $\frac{1}{3}Al^{3+}$ to form an Al complex with NH_4^+]

Referring to the above methods, mineral drugs containing Al^{3+} can be identified, for example, alum [the main component is $KAl(SO_4)_2 \cdot 12H_2O$], etc.

1.3 Centrifuge Tube Test

The method of placing the test solution and the reagent in a centrifuge tube to perform a precipitation reaction and allowing the precipitate to be centrifuged out on a centrifuge is called a centrifuge tube test method. When starting the centrifuge, pay attention to the stability of the centrifuge. The centrifuge tube should be placed in the sleeve of the centrifuge to achieve symmetry. If necessary, water should be added to the empty centrifuge tube for symmetry. After centrifugation, the precipitate is deposited on the tip of the centrifuge tube by centrifugation. The centrifuged liquid can be sucked out with a capillary tube, leaving the precipitate in the centrifuge tube.

(1) Ag^+: Take a little Ag^+ solution, add a few drops of dilute hydrochloric acid to observe the formation of white curd-like precipitate, then centrifuge, discard the supernatant. Drop the ammonia solution on the precipitate, observe the dissolution of precipitation. Write down the ion chemical reaction.

(2) Sb^{3+}: Take a small amount of Sb^{3+} solution, add a few drops of dilute hydrochloric acid to acidify, pass hydrogen sulfide gas, observe the formation of the precipitate, centrifuge, discard the supernatant, drop the sodium sulfide solution on the precipitate, and observe the dissolution of the precipitate. Write down the ion reaction.

1.4 Titration Test Method on Paper

The test solution and reagent are dropped on the filter paper for precipitation. Due to the capillary action of the filter paper, except the precipitation, other ions will diffuse uniformly outside the precipitation area. The phenomena of precipitation formation should be obvious. This method is called titration test method on paper.

Ba^{2+} salt: Drop a small amount of Ba^{2+} salt solution on the filter paper, add rhodizonic acid sodium reagent solution, observe the formation of the precipitate and the color change of the filter paper after treatment with dilute hydrochloric acid.

1.5 Gas Chamber Test

The gas chamber is composed of two small watch glasses which are put together. The upper one can be slightly smaller and wiped dry. Wet the test paper and stick it to the center of the concave surface of the upper watch glass which then covers on watch glass below. If necessary, it can be placed on a small beaker and heated with a steam bath. After the reaction occurs, observe the color change of the filter paper. This method of identifying the presence of ions is called the gas chamber test.

(1) NH_4^+ salt: take a small amount of NH_4^+ salt solid, add an excess of NaOH solution, heat, and the gas generated by the decomposition will cause stains on the filter paper impregnated with Nesler reagent. Observe the formation of the stain and write the ion reaction formula.

(2) As^{3+} salt: put a little As^{3+} solid on a watch glass, add a small amount of arsenic-free zinc powder and dilute sulfuric acid, and stick a piece of filter paper with silver nitrate solution on the upper watch glass of the gas chamber. After a few minutes, observe the formation of stains on the filter paper.

(3) SO_4^{2-} salt: put a little SO_4^{2-} salt solid on a watch glass, add hydrochloric acid to generate a gas which can make the filter paper moistened with solution of mercury (I) nitrate to produce stains. Observe the occurrence of stains.

1.6 Flame Test

The platinum wire (or nickel-chromium wire) is made into a ring, a few drops of concentrated hydrochloric acid are placed on the drip plate, then insert the metal ring into the hydrochloric acid and soak, and then burn in the flame. Repeated these steps several times until the flame was not stained, indicating that the wire had been treated clean. Using this clean metal wire to dip the test solution and burn it in an oxidizing flame, identify inorganic ions by characteristic color of flame. This method is called flame test.

(1) Sodium salt: bright yellow in flame.

(2) Potassium salt: purple in flame (When sodium salts are mixed, blue cobalt glass can be used for perspective.)

(3) Calcium salt: brick red in flame.

(4) Barium salt: yellow-green in flame.

Record the flame color of each salt in the flame, and the wire should be cleaned by the method described above before each experiment. Never place the platinum wire in a reducing flame to burn it, otherwise platinum carbide will be generated, and the platinum wire will be brittle.

2. Identification of Unknowns

Receive an unknown solution, which may contain Na^+, K^+, NH_4^+, Mg^{2+}, Ca^{2+}, Ba^{2+}, Cl^-, Br^-, and I^-. With reference to the above experiments, develop your own analysis steps to determine which ions are contained in the unknown solution.

3. Identification of Mineral Drugs

Two unknown mineral drugs were collected, including Na^+, K^+, Al^{3+}, Cl^-, NO_3^-, and SO_4^{2-}. Please design an experimental scheme to identify the unknown mineral drugs.

Questions

(1) What operating techniques are used to identify common inorganic ions?

(2) What are the prerequisites for carrying out the "general identification reaction" of ions?

(3) An unknown solution contains Br^- but no Cl^-. When HNO_3 solution is added to a solution containing 2mol/L $NH_3 \cdot H_2O$ used to treat a AgX (X is halide) precipitate, turbidity is observed. Try to explain this phenomenon.

附 录

附录1 实验室常用试剂的配制

一、常用酸碱溶液

试剂名称	物质的量浓度 (mol/L)	质量分数 (%)	密度 (20℃) (g/ml)	配制方法
浓盐酸 (HCl)	12	37.23	1.19	
稀盐酸 (HCl)	6	20.4	1.10	浓盐酸 496ml,用水稀释至 1L
稀盐酸 (HCl)	2	7	1.03	浓盐酸 167ml,用水稀释至 1L
浓硝酸 (HNO_3)	15	68	1.40	
稀硝酸 (HNO_3)	6	32	1.20	浓硝酸 375ml,用水稀释至 1L
稀硝酸 (HNO_3)	2	12	1.07	浓硝酸 125ml,用水稀释至 1L
浓硫酸 (H_2SO_4)	18	98	1.84	
稀硫酸 (H_2SO_4)	6	44	1.34	浓硫酸 334ml,慢慢加到 600ml 水中,并不断搅拌,再用水稀释至 1L
稀硫酸 (H_2SO_4)	2	9	1.06	浓硫酸 111ml,慢慢加到 800ml 水中,并不断搅拌,再用水稀释至 1L
冰醋酸 (HAc)	17	99	1.05	
稀乙酸 (HAc)	5	30	1.04	冰醋酸 295ml,用水稀释至 1L
稀乙酸 (HAc)	2	12	1.02	冰醋酸 118ml,用水稀释至 1L
浓磷酸 (H_3PO_4)	14.7	85	1.69	
稀磷酸 (H_3PO_4)	1	9	1.05	浓磷酸 68ml,用水稀释至 1L
浓氢氟酸 (HF)	23	40	1.13	
氢溴酸 (HBr)	7	40	1.38	
氢碘酸 (HI)	7.5	57	1.70	
浓高氯酸 ($HClO_4$)	11.6	70	1.67	
稀高氯酸 ($HClO_4$)	2	19	1.12	浓高氯酸 172ml,用水稀释至 1L

续表

试剂名称	物质的量浓度 (mol/L)	质量分数 (%)	密度(20℃) (g/ml)	配制方法
浓氨水 ($NH_3 \cdot H_2O$)	14.8	25~27	0.90	
稀氨水 ($NH_3 \cdot H_2O$)	6	10	0.96	浓氨水 400ml，用水稀释至 1L
氢氧化钾 (KOH)	6	25	1.25	337g KOH 溶于适量水中稀释至 1L
浓氢氧化钠 (NaOH)	6	20	1.22	240g NaOH 溶于适量水中稀释至 1L
稀氢氧化钠 (NaOH)	2	8	1.09	80g NaOH 溶于适量水中稀释至 1L
氢氧化钙 [$Ca(OH)_2$]	6	0.025		饱和溶液
氢氧化钡 [$Ba(OH)_2$]	6	0.2		饱和溶液

二、实验室常用试剂

试剂名称	浓度	配制方法
醋酸钠 ($NaAc \cdot 3H_2O$)	1 mol/L	溶解 136g $NaAc \cdot 3H_2O$ 于水中，稀释至 1L
醋酸铅 [$Pb(Ac)_2 \cdot 3H_2O$]	1 mol/L	溶解 379g $Pb(Ac)_2 \cdot 3H_2O$ 于水中，稀释至 1L
碘化钾 (KI)	1 mol/L	溶解 166g KI 于水中，稀释至 1L
碘溶液	0.01 mol/L	溶 1.3g 碘和 3g KI 于尽可能少量的水中，稀释至 1L
淀粉溶液	1%	将 1g 淀粉和少量冷水调成糊状，倒入 100ml 沸水中，煮沸后冷却即可
丁二酮肟	1%	溶解 1g 丁二酮肟于 100ml 95% 乙醇中
高锰酸钾 ($KMnO_4$)	0.03%	溶解 0.3g $KMnO_4$ 于水中，稀释至 1L
高锰酸钾 ($KMnO_4$)	0.1 mol/L	溶解 16g $KMnO_4$ 于水中，稀释至 1L
高锰酸钾 ($KMnO_4$)	饱和	溶解 70g $KMnO_4$ 于水中，稀释至 1L
铬酸钾 (K_2CrO_4)	1 mol/L	溶解 194g K_2CrO_4 于水中，稀释至 1L
过氧化氢 (H_2O_2)	3%	量取 100ml 30% H_2O_2 溶液，用水稀释至 1L
邻二氮菲	0.5%	0.5g $FeSO_4$ 溶于 100ml 水，加 2 滴硫酸与 0.5g 邻二氮菲，摇匀
磷酸氢二钠 ($Na_2HPO_4 \cdot 12H_2O$)	0.1 mol/L	溶解 35.82g $Na_2HPO_4 \cdot 12H_2O$ 于水中，稀释至 1L
硫代硫酸钠 ($Na_2S_2O_3 \cdot 5H_2O$)	0.1 mol/L	溶解 24.82g $Na_2S_2O_3 \cdot 5H_2O$ 于水中，稀释至 1L
硫化钠 (Na_2S)	1 mol/L	溶解 240g $Na_2S \cdot 9H_2O$ 和 40g NaOH 于水中，稀释至 1L
硫酸铵 [$(NH_4)_2SO_4$]	1 mol/L	溶解 132g $(NH_4)_2SO_4$ 于水中，稀释至 1L
硫酸锌 ($ZnSO_4 \cdot 7H_2O$)	0.1 mol/L	溶解 28.7g $ZnSO_4 \cdot 7H_2O$ 于水中，稀释至 1L
硫酸锌 ($ZnSO_4 \cdot 7H_2O$)	饱和	溶解约 900g $ZnSO_4 \cdot 7H_2O$ 于水中，稀释至 1L
硫酸亚铁 ($FeSO_4 \cdot 7H_2O$)	1 mol/L	用适量稀硫酸溶解 278g $FeSO_4 \cdot 7H_2O$，加水稀释至 1L
铝试剂	1 mol/L	1g 铝试剂溶于 1L 水中
氯化铵 (NH_4Cl)	1 mol/L	溶解 53.5g NH_4Cl 于水中，稀释至 1L
氯化钡 ($BaCl_2 \cdot 2H_2O$)	0.1 mol/L	溶解 24.4g $BaCl_2 \cdot 2H_2O$ 于水中，稀释至 1L
氯化钡 ($BaCl_2 \cdot 2H_2O$)	25%	溶解 250g $BaCl_2 \cdot 2H_2O$ 于水中，稀释至 1L
氯化钾 (KCl)	1 mol/L	溶解 74.5g KCl 于水中，稀释至 1L
氯化铁 ($FeCl_3 \cdot 6H_2O$)	1 mol/L	溶解 270g $FeCl_3 \cdot 6H_2O$ 于适量浓盐酸中，加水稀释至 1L

续表

试剂名称	浓度	配制方法
氯化亚锡 ($SnCl_2 \cdot 2H_2O$)	0.1 mol/L	溶解 22.5g $SnCl_2 \cdot 2H_2O$ 于 150ml 浓盐酸中，加水稀释至 1L，加入数粒锡粒，以防氧化
氯水	饱和	通氯气于水中至饱和为止
镁试剂（对 – 硝基苯偶氮 – 间苯二酚）		溶解 0.01g 镁试剂于 1L 的 1 mol/L NaOH 溶液中
奈斯勒试剂		将 115g HgI_2 和 80g KI 溶解于 50ml 水中，加入 500ml 6 mol/L 的 NaOH 溶液，混匀、静置，吸取上层清液，储于棕色瓶中
碳酸钠 (Na_2CO_3)	1 mol/L	溶解 106g Na_2CO_3 于水中，稀释至 1L
铁氰化钾 [$K_3Fe(CN)_6$]	1 mol/L	溶解 329g $K_3Fe(CN)_6$ 于水中，稀释至 1L
硝酸铵 (NH_4NO_3)	1 mol/L	溶解 80g NH_4NO_3 于水中，稀释至 1L
硝酸银 ($AgNO_3$)	0.1 mol/L	用水溶解 17.0g $AgNO_3$，加水稀释至 1L
溴水	饱和	在水中滴入液溴至饱和为止
亚铁氰化钾 [$K_4Fe(CN)_6 \cdot 3H_2O$]	1 mol/L	溶解 422.4g $K_4Fe(CN)_6 \cdot 3H_2O$ 于水中，稀释至 1L

三、常用缓冲溶液

缓冲溶液组成	pH	密度 (20℃)(g/ml)
氨基乙酸 –HCl	2.30	取氨基乙酸 150g 溶于 500ml 水中，加浓 HCl 80ml，加水稀释至 1L
磷酸 – 柠檬酸盐	2.50	取 $Na_2HPO_4 \cdot 12H_2O$ 113g 溶于 200ml 水中，加柠檬酸 387g，溶解，过滤，加水稀释至 1L
一氯乙酸 –NaOH	2.80	取 500g 一氯乙酸溶于 200ml 水中，加 NaOH 40g 溶解后，加水稀释至 1L
甲酸 –NaOH	4.00	将 95g 甲酸和 40g NaOH 溶于 500ml 水中，加水稀释至 1L
NaAc–HAc	4.50	取 64g $NaAc \cdot 3H_2O$ 溶于适量水中，加 6mol/L HAc 136ml，加水稀释至 1L
NaAc–HAc	5.00	取 100g $NaAc \cdot 3H_2O$ 溶于适量水中，加 6mol/L HAc 68ml，加水稀释至 1L
NaAc–HAc	5.70	取 200g $NaAc \cdot 3H_2O$ 溶于适量水中，加 6mol/L HAc 26ml，加水稀释至 1L
NH_4Cl–NH_3	8.00	取 100g NH_4Cl 溶于适量水中，加浓氨水 7.0ml，加水稀释至 1L
NH_4Cl–NH_3	8.50	取 140g NH_4Cl 溶于适量水中，加浓氨水 8.8ml，加水稀释至 500ml
NH_4Cl–NH_3	9.00	取 70g NH_4Cl 溶于适量水中，加浓氨水 48ml，加水稀释至 1L
NH_4Cl–NH_3	9.50	取 54g NH_4Cl 溶于适量水中，加浓氨水 126ml，加水稀释至 1L
NH_4Cl–NH_3	10.00	取 54g NH_4Cl 溶于适量水中，加浓氨水 350ml，加水稀释至 1L

附录 2 常用的酸碱指示剂

指示剂名称	变色 pH 范围	颜色变化		配制方法
		酸色	碱色	
百里酚蓝	2.0~2.8	红	黄	0.1g 百里酚蓝溶于 100ml 20% 乙醇
甲基黄	2.9~4.0	红	黄	0.1g 甲基黄溶于 100ml 90% 乙醇
甲基橙	3.1~4.4	红	黄	0.1g 甲基橙溶于 100ml 热水
溴酚蓝	3.0~4.6	黄	紫蓝	0.1g 溴酚蓝溶于 100ml 20% 乙醇
刚果红	3.0~5.2	蓝紫	红	0.1g 刚果红溶于 100ml 水中
溴甲酚绿	3.8~5.4	黄	蓝	0.1g 溴甲酚绿溶于 100ml 20% 乙醇
甲基红	4.4~6.2	红	黄	0.1g 甲基红溶于 100ml 20% 乙醇
石蕊	4.5~8.3	红	蓝	0.1g 石蕊溶于 100ml 乙醇
溴百里酚蓝	6.0~7.6	黄	蓝	0.1g 溴百里酚蓝溶于 100ml 20% 乙醇
酚红	6.4~8.0	黄	红	0.1g 酚红溶于 100ml 20% 乙醇
中性红	6.8~8.0	红	黄	0.1g 中性红溶于 100ml 60% 乙醇
酚酞	8.0~10.0	无色	红	0.1g 酚酞溶于 100ml 乙醇
百里酚酞	9.4~10.6	无色	蓝	0.1g 百里酚酞溶于 100ml 90% 乙醇
茜素黄	10.1~12.1	黄	紫	0.1g 茜素黄溶于 100ml 水中
靛蓝胭脂红	11.6~14.0	蓝	黄	0.25g 靛蓝胭脂红溶于 100ml 50% 乙醇
1,3,5- 三硝基苯	12.2~14.0	无色	蓝	0.18g 1,3,5- 三硝基苯溶于 100ml 90% 乙醇

附录3 常见离子和化合物的颜色

一、常见离子的颜色

离子	颜色	离子	颜色	离子	颜色	离子	颜色
$[Ag(NH_3)_2]^+$	无色	$[Ag(S_2O_3)_2]^{3-}$	无色	Co^{2+}	桃红	$[Co(CN)_6]^{3-}$	紫色
$[Co(NH_3)_6]^{2+}$	橙色	$[Co(NH_3)_6]^{3+}$	酒红	$[Co(NO_2)_6]^{3-}$	黄色	CrO_4^{2-}	橘黄
$Cr_2O_7^{2-}$	橘红	$[Cu(NH_3)_4]^{2+}$	深蓝色	$[Cu(OH)_4]^{2-}$	蓝色	$[CuCl_4]^{2-}$	黄色
$[Fe(CN)_6]^{3-}$	无色	$[Fe(CN)_6]^{4-}$	黄色	$[HgCl_4]^{2-}$	无色	$[HgI_4]^{2-}$	无色
$[Ni(CN)_4]^{2-}$	无色	$[Ni(NH_3)_6]^{2+}$	紫色	$[Zn(NH_3)_4]^{2+}$	无色	Al^{3+}	无色
Ca^{2+}	无色	Cr^{3+}	绿色	CrO_2^-	亮绿色	Cu^+	蓝色
Fe^{2+}	绿色	Fe^{3+}	浅紫	K^+	无色	Mg^{2+}	无色
Mn^{2+}	浅粉色	MnO_4^-	紫色	MnO_4^{2-}	绿色	Na^+	无色
NH_4^+	无色	Ni^{2+}	绿色	SCN^-	无色	Br^-	无色

二、常见化合物的颜色

化合物	颜色	化合物	颜色	化合物	颜色	化合物	颜色
$(NH_4)_2HPO_4(s)$	白色	$(NH_4)_2S_2O_8(s)$	白色	$(NH_4)_2SO_4(s)$	白色	$(NH_4)H_2PO_4(s)$	白色
$Ag_2Cr_2O_7(s)$	深红色	$Ag_2CrO_4(s)$	砖红	$Ag_2O(s)$	棕黑	$Ag_2S(s)$	灰黑
$AgBr(s)$	淡黄	$AgCl(s)$	白色	$AgI(s)$	黄色	$AgNO_3(s)$	无色
$AgSCN(s)$	无色	$Al(OH)_3(s)$	白色	$As_2O_3(s)$	白色	$Ba(OH)_2(s)$	白色
$BaCl_2(s)$	白色	$BaCrO_4(s)$	黄色	$BaSO_4(s)$	白色	$Ca(ClO)_2(s)$	白色
$Ca(H_2PO_4)_2(s)$	无色	$Ca_3(PO_4)_2(s)$	白色	$CaCl_2(s)$	白色	$CaCO_3(s)$	白色
$CaCrO_4(s)$	黄色	$CaHPO_4(s)$	白色	$CaSO_4(s)$	白色	$CdCl_2(s)$	白色或无
$CdCl_2 \cdot 6H_2O(s)$	粉色	$CdS(s)$	淡黄	$CoCl_2(s)$	蓝色	$CoCl_2 \cdot 6H_2O(s)$	蓝色
$CuCl_2 \cdot 2H_2O(s)$	蓝色	$CuBr_2(s)$	黑紫色	$CoSO_4(s)$	红色	$Cr(OH)_3(s)$	灰绿
$Cr_2O_3(s)$	亮绿	$CrCl_3(s)$	暗绿	$Cu(OH)_2(s)$	蓝色	$Cu_2O(s)$	红棕色
$Cu_2S(s)$	蓝~灰黑	$CuO(s)$	黑色	$CuS(s)$	黑色	$CuSO_4(s)$	灰白色
$CuSO_4 \cdot 5H_2O(s)$	蓝色	$Fe(OH)_3(s)$	红~棕色	$Fe_2O_3(s)$	红棕色	$Fe_2S_3(s)$	黄绿色
$FeCl_2(s)$	灰绿	$FeCl_3(s)$	暗红色	$FeS(s)$	黑色	$FeSO_4 \cdot 7H_2O(s)$	蓝绿色
$H_2O_2(l)$	无色	$Hg(NO_3)_2(s)$	无色	$Hg(NO_3)_2 \cdot H_2O(s)$	无或微黄	$Hg_2Cl_2(s)$	白色
$Hg_2I_2(s)$	亮黄	$HgCl_2(s)$	白色	$HgI_2(s)$	猩红	$HgNH_2Cl(s)$	白色

续表

化合物	颜色	化合物	颜色	化合物	颜色	化合物	颜色
HgO(s)	亮红	HgS(s) 六方	红色	HgS(s) 立方	黑色	$K_2Cr_2O_7$(s)	橘红色
K_2CrO_4(s)	柠檬黄	K_2MnO_4(s)	绿色	$K_2S_2O_3$(s)	无色	K_2SO_3(s)	白色
K_2SO_4(s)	无或白色	$K_3Fe(CN)_6$(s)	宝石红	$K_4Fe(CN)_6 \cdot 3H_2O$(s)	黄色	KBr(s)	白色
KCl(s)	无或白色	KCN(s)	白色	KI(s)	白色	KIO_3(s)	白色
$KMnO_4$(s)	紫色	KNO_2(s)	白色或微黄	KNO_3(s)	无色	KOH(s)	白色
KSCN(s)	无色	$MgSO_4 \cdot 7H_2O$(s)	白色	$MnCl_2$(s)	淡红色	MnO_2(s)	紫黑
MnS(s)	浅红	$MnSO_4$(s)	淡红	$Na_2B_4O_7$(s)	白色	Na_2CO_3(s)	白色
$Na_2CO_3 \cdot 10H_2O$(s)	无色	$Na_2Cr_2O_7$(s)	橙红色	Na_2CrO_4(s)	黄色	Na_2HPO_4(s)	无色
Na_2S(s)	无色	$Na_2S_2O_3$(s)	白色	Na_2SO_3(s)	白色	Na_2SO_4(s)	无色
Na_2SO_4(s) $10H_2O$(s)	无色	Na_3PO_4(s)	无色	NaAc(s)	白色	NaCl(s)	白色
NaF(s)	无色	NaH_2PO_4(s)	无色	$NaHCO_3$(s)	白色	NaI(s)	白色
NH_4Br(s)	白色	NH_4Cl(s)	白色	NH_4F(s)	白色	NH_4NO_3(s)	无或白色
NH_4SCN(s)	无色	$Ni(OH)_2$(s)	苹果绿	$NiCl_2$(s)	绿色	NiS(s)	黑色
$NiSO_4$(s)	翠绿	$Pb(Ac)_2$(s)	无或白色	$Pb(NO_3)_2$(s)	白色或无	$PbCl_2$(s)	色白或无
$PbCrO_4$(s)	橙黄色	PbO_2(s)	深棕色	PbS(s)	黑色	$PbSO_4$(s)	白色
$SnCl_2$(s)	白色	$SnCl_4$(s)	无色	SnS(s)	棕色	ZnS(s)	白或淡黄

附录4 常见阴、阳离子鉴定一览表

离子	试剂及鉴定方法	现象	备注
Br^-	取2滴试液，加入数滴CCl_4，滴入氯水，振荡	有机层显红棕色或金黄色	
Cl^-	与银氨溶液和HNO_3反应	有白色沉淀析出	
CO_3^{2-}	使试液与$Ba(OH)_2$作用	有白色浑浊出现	
I^-	取2滴试液，加入数滴CCl_4，滴入氯水，振荡	有机层显紫色	
NO_2^-	取1滴试液加6mol/L HAc酸化，加1滴对氨基苯磺酸，1滴α-萘胺	溶液显红紫色	HAc介质
NO_3^-	在小试管中滴加10滴饱和$FeSO_4$溶液，5滴试液，然后斜持试管，沿着管壁慢慢滴加浓H_2SO_4	溶液分层，在两层接触界面有棕色环	硫酸介质
PO_4^{3-}	取2滴试液，加入8~10滴钼酸铵试剂，用玻璃棒摩擦器壁	有黄色沉淀生成	硝酸介质
S^{2-}	取3滴试液，加稀H_2SO_4酸化，用$Pb(Ac)_2$试纸检验放出的气体	试纸变黑	酸性介质
$S_2O_3^{2-}$	取2滴试液，加2滴2mol/L HCl溶液，加热	有白色浑浊出现	酸性介质
SO_3^{2-}	取1滴$ZnSO_4$饱和溶液，加1滴$K_4[Fe(CN)_6]$于点滴板中，继续加入1滴$Na_2[Fe(CN)_5NO]$，1滴试液	先有白色沉淀生成后转化为红色沉淀	氨水介质
SO_4^{2-}	试液用6mol/L HCl酸化，加2滴0.5mol/L $BaCl_2$溶液	有白色沉淀析出	酸性介质
Ag^+	取2滴试液，加2滴2mol/L HCl溶液，搅动，水浴加热，离心分离，沉淀加氨水，再加6mol/L HNO_3酸化	有白色沉淀析出，后溶解，加酸后又有白色沉淀生成	酸性介质
Al^{3+}	取1滴试液，加2~3滴水，加2滴3mol/L NH_4Ac，2滴铝试剂，搅拌，微热片刻，加6mol/L氨水至碱性	有红色沉淀生成且不消失	HAc-NH_4Ac介质
Ba^{2+}	取2滴试液，加1滴0.1mol/L K_2CrO_4溶液	有黄色沉淀生成	HAc-NH_4Ac介质
Ca^{2+}	取2滴试液，滴加饱和$(NH_4)_2C_2O_4$溶液	有白色沉淀生成	氨水介质
Co^{2+}	取2滴试液，加饱和NH_4SCN溶液，加5~6滴戊醇溶液，振荡，静置	有机层呈蓝绿色	中性介质
Cr^{3+}	取3滴试液，加6mol/L NaOH溶液至澄清，搅动后加4滴3%的H_2O_2，水浴加热，冷却，6mol/L HAc酸化，加2滴0.1mol/L $Pb(NO_3)_2$溶液	溶液颜色由绿变黄，后有黄色沉淀生成	强碱性介质
Cu^{2+}	取1滴试液，加1滴6mol/L HAc酸化，加1滴$K_4[Fe(CN)_6]$溶液	有红棕色沉淀生成	中性或弱碱性介质
Fe^{2+}	取1滴试液滴在点滴板上，加1滴$K_3[Fe(CN)_6]$溶液	有蓝色沉淀生成	酸性介质
Fe^{3+}	取1滴试液滴在点滴板上，加1滴$K_4[Fe(CN)_6]$溶液	有蓝色沉淀生成	酸性介质
Hg^{2+}	取2滴试液，加1mol/L KI溶液，使生成沉淀后又溶解，加2滴$KI-Na_2SO_3$溶液，2~3滴Cu^{2+}溶液	有橘黄色沉淀生成	
K^+	取2滴试液，加3滴$Na_3[Co(NO_2)_6]$溶液，放置片刻	有黄色沉淀生成	中性或微酸性介质

续表

离子	试剂及鉴定方法	现象	备注
Mg^{2+}	取 2 滴试液，加 2 滴 2mol/L NaOH 溶液，1 滴镁试剂	有天蓝色沉淀生成	碱性介质
Mn^{2+}	取 1 滴试液，加 10 滴水，5 滴 2mol/L HNO_3 溶液，然后加固体 $NaBiO_3$，搅拌，水浴加热	溶液呈紫色	酸性介质
Na^+	取 2 滴试液，加 8 滴醋酸铀酰试剂，放置数分钟，用玻璃棒摩擦器壁	有淡黄色的晶状沉淀出现	中性或 HAc 酸性介质
NH_4^+	取 1 滴试液，放在点滴板的圆孔中，加 2 滴奈氏试剂	有红棕色沉淀生成	碱性介质
Ni^{2+}	取 1 滴试液滴在点滴板上，加 1 滴 6mol/L 氨水，加 1 滴丁二酮肟	在凹槽四周有红色沉淀生成	氨性介质
Pb^{2+}	取 2 滴试液，加 2 滴 0.1mol/L K_2CrO_4 溶液	有黄色沉淀生成	HAc 介质
Sn^{2+}	取 2 滴试液，加 1 滴 0.1mol/L $HgCl_2$ 溶液	有白色沉淀生成	酸性介质
Zn^{2+}	取 2 滴试液，用 2mol/L HAc 酸化，加等体积的 $(NH_4)_2Hg(SCN)_4$ 溶液，摩擦器壁	有白色沉淀生成	中性或微酸性介质

参考文献

[1] 铁步荣，杨怀霞. 无机化学实验 [M]. 北京：中国中医药出版社，2016.

[2] 吴巧凤，刘幸平. 无机化学实验 [M]. 北京：人民卫生出版社，2012.

[3] 曹凤歧. 无机化学实验与指导（Experiment and Guide for Inorganic Chemistry）[M]. 2版. 北京：中国医药科技出版社，2006.

[4] 杨怀霞，吴培云. 无机化学实验 [M]. 2版. 北京：中国医药科技出版社，2018.

[5] 李发美. 分析化学实验指导 [M]. 2版. 北京：人民卫生出版社，2007.

[6] 严拯宇，杜迎翔. 分析化学实验与指导（Experiment and Guide for Analytical Chemistry）[M]. 3版. 北京：中国医药科技出版社，2015.

[7] 大连理工大学无机化学教研室编. 无机化学实验 [M]. 2版. 北京：高等教育出版社，2004.

[8] 天津大学无机化学教研室编. 无机化学实验 [M]. 修订版，天津：天津大学出版社，2003.

[9] 童国秀. 无机化学实验 [M]. 英汉双语版. 北京：科学出版社，2019.

[10] 杨芳，郑文杰. 无机化学实验 [M]. 中英双语版. 北京：化学工业出版社，2014.

[11] 王秋长，赵鸿喜，张守民，等. 基础化学实验 [M]. 北京：科学出版社，2003.

[12] Gary L. Miessler, Donald A. Tarr. 无机化学 [M]. 3版. 影印本. 北京：高等教育出版社，2006.

[13] Ralph H. Petrucci, William S. Harwood, F. Geoffrey Herring. 普通化学原理与应用 [M]. 8版. 影印本. 北京：高等教育出版社，2004.

[14] Kenneth W. Whitten, Raymond E. Davis, M. Larry Peck, George G. Stanley. Chemistry[M]. 9rd Ed. USA. BROOKS/COLE, 2010.